Breathless Sleep ... *no more*

A compelling case study

Breathless Sleep ... *no more*

A compelling case study

Paul Rodriguez

Cover Design by Megan Slattery Design

A Catalogue-in-Publication is available from the National Library of Australia.

Creator: Rodriguez, Paul, author.

Title: Breathless sleep ... no more : a compelling case study/Paul Rodriguez.

ISBN: 9780994169402 (paperback)

Subjects: Sleep apnoea syndromes--Treatment.

Sleep disorders--Treatment.

Apnoea.

Dewey Number: 616.209

To the memory of Glenda, my mother, (1935–2006) who struggled in ignorance with the debilitating effects of sleep apnoea and was denied the opportunity to address the causes of her illness.

I would like to thank the following people who generously gave me their assistance and guidance in writing this book, namely, Deb Harris, Michael Pardy, Dr Natalie Bird, Paul O'Connell and Carolyn Barnes. Finally I express my indebtedness to Claire Benton for her invaluable mentoring and instruction in the Buteyko breathing method.

In this second edition I wish to acknowledge Patrick McKeown for his valuable assistance regarding that part of Chapter 7 devoted to the physiological benefits of nitric oxide. I also wish to acknowledge Dr Lauren Boundy's valuable assistance regarding Chapter 6 devoted to my tongue tie release. Finally I thank Bridget Ingle for her perception, her enthusiasm and her generous time given while tongue tutoring before and after my tongue tie release.

Contents

Foreword to 1st edition

I was pleasantly surprised to receive a phone call from Paul Rodriguez advising that he had just written a book about overcoming his debilitating sleep apnoea with the Buteyko Breathing Method. He also advised that he had been introduced to Buteyko Breathing by attending a seminar I conducted at the Council of Adult Education in Melbourne in 2008, and asked if I was interested in writing the foreword for his book. I was delighted to accept.

Paul outlines in detail his struggle with poor sleep over many years and the effect this had on several aspects of his life including weight gain, stress, depression, energy levels and relationships. He was eventually diagnosed with sleep apnoea and initially tried to get relief by using a CPAP machine. His honesty and openness about his problems will strike a chord with many thousands of people.

He explains how by changing his breathing he quickly started to improve. He provides a clear outline of the science behind Buteyko Breathing and how and why his health improved. The references and technical data relating to his case will make it easy for people to understand the results he achieved with Buteyko, which at first glance often seem too good to be true.

Having taught the Buteyko Breathing Method to more than 8,000 people over the last 20 years, I have received many hundreds of testimonials and case studies from happy clients.

But no-one has taken such time and effort to share their experience for the benefit of others like Paul Rodriguez.

I congratulate him and am sure you will be inspired by his book.

Paul O'Connell
BSc, Dip Ed, MBA
Chief Executive Officer
Buteyko Institute of Breathing & Health

Foreword to 2nd edition

To wake up alert and refreshed, there are two simple strategies contained in this book which you can immediately apply. The first is to breathe silently and the second is to breathe only through your nose during the daytime and at sleep.

Both strategies are logical, have no side effects and are supported by decades of research.

If you were to witness someone snore or stop breathing during sleep, you would notice that their breathing is hard and often through the mouth. As long as this breathing pattern continues, their sleep quality is very likely to remain poor. If you wake up with a dry mouth in the morning, your sleep quality can certainly be improved.

The benefits of breathing silently and through the nose are numerous. The nose has its own supply of the gas nitric oxide. As each breath is drawn through the nose, nitric oxide signals the muscles of the throat to stay wide and open. Without this, the throat can collapse, a person can stop breathing, and their sleep and health can adversely be affected. In addition, the nose is directly linked to our main breathing muscle, the diaphragm. Using the diaphragm to breathe improves lung volume and this helps ensure that the muscles of the throat work harder to stay open during sleep. Breathing through the nose is imperative to harness this constant communication between the nose, diaphragm and throat.

Another consideration is breathing volume and the negative pressure created in the upper airways during breathing. The harder we breathe, the greater the negative pressure as air is drawn into the lungs. Imagine the upper airways of the throat as a collapsible paper straw. If one breathes hard through the straw, then increased negative pressure will cause the sides of the straw to be drawn inwards thereby causing collapse. If on the other hand, breathing is light and through the nose, there is less negative pressure and therefore less collapse.

The ideal resting position of the tongue is in the roof of the mouth. During the day and during sleep, three quarters of the tongue should be against the palate. However, if the person breathes through their mouth or if their tongue is tied too tightly to the floor of the mouth, the tongue is prevented from resting in the ideal position. When the tongue rests midway or on the floor of the mouth, the back of the tongue is encroaching on the space of the throat. This reduces the size of the upper airway making it more liable to collapse.

In sixteenth century Europe, it was a practice of midwives to check whether a new born baby had free movement of their tongue. If their tongue was held too tightly to the floor of the mouth, the midwife clipped the piece of string holding the tongue with their finger nail. This procedure was a case of life or death for the baby, as only with a free tongue could the baby latch to the breast for feeding. While breast feeding provides the baby with good nutrition, it also helps to ensure normal development of the jaws and face which is necessary to support nasal breathing. In today's society, babies with tongue tie often fall between the cracks of modern medicine. These infants can experience difficulty feeding from their mother. They have an increased tendency towards mouth breathing which reinforces poor development of their airway. It is tragic to note that up to 50% of studied children are persistent mouth breathers and this sets the child up for lifelong sleep problems, something that could be avoided with a little awareness. Addressing tongue ties in young babies and ensuring that the child breathes through his or her nose is essential for the future of a thriving society.

In his quest to improve his sleep, Paul Rodriguez had his tongue clipped. With freer movement of the tongue, Paul is able to maintain the correct tongue posture and make room in his airway for breathing.

The nose is designed by nature to ensure that breathing is efficient. Mouth breathing can cause too much carbon dioxide to leave the body resulting in blood pH becoming too alkaline. As carbon dioxide is the primary regulator of blood pH, removing too much by breathing hard will reduce the signals from the brain to breathe. When the brain does not send the message to breathe during sleep, this is called a central sleep apnoea. Sleep apnoea can be a combination of the airways collapsing during sleep to cause obstructive sleep apnoea along with reduced signals by the brain to breathe causing central sleep apnoea.

Another purpose of the nose is to condition incoming air to the correct moisture levels and temperature for the lungs. The mouth is not nearly as effective as the nose in moistening incoming air. Breathing through the mouth causes a drying of the airways making them sticky. When sticky airways collapse, a person stops breathing for a longer period of time. This increases the severity of sleep apnoea, as blood oxygen saturation drops severely. Mouth breathing also causes trauma to the upper airways, literally sucking moisture and contributing to inflammation. When the airways are inflamed, they become narrow, leading to breathing difficulties.

Paul Rodriguez is not a sleep doctor. Instead he is office based and performs tasks which require good quality sleep to stay focused. Like many of us, Paul suffered from sleep problems and this affected his work and quality of life. Breathless Sleep ... *no more* is not just Paul's account of his journey to getting a good night's sleep. It is an account of the tools to help you get a good night's sleep.

Read Paul's sleep studies and observe what happened when he made simple changes to his breathing. I believe the best solution to health problems is learning and adopting practices that truly help us. Breathless Sleep ... *no more* does just that.

Patrick McKeown
Author, The Oxygen Advantage

Introduction

Why I am compelled to write this book

I am fifty-eight years old. I was afflicted with sleep apnoea for twenty years or so. But I have only come to know this fact in the last ten years.

It has been found that moderate to severe obstructive sleep apnoea is associated with an increased risk of death from any cause in middle-aged adults, especially men. In the USA, in 2009, more than 12 million adults (or more than 4% of the population) were believed to have sleep apnoea, and most of them were neither diagnosed nor treated.[1]

Experts estimate that sleep apnoea affects 100 million people worldwide. Approximately 80% of those people are currently undiagnosed. According to a 2009 study in the USA it is estimated that 7% of Americans suffer from sleep apnoea with 9% of male adults and 4% of female adults affected by moderate to severe sleep apnoea. Children comprise 2% of those affected by sleep apnoea, most of them being less than one year old.[2]

In 2017 in the USA some forty-seven million adults, according to the National Sleep Foundation, do not get a restorative night's sleep. In the workplace, sleep deprivation results in injuries and decreased productivity, which is thought to cost the USA 18 billion dollars each year."[3]

In 2012 in Australia more than 1.5 million adults suffered from sleep disorders and more than 1 million (or 4.7% of the population) specifically suffered from obstructive sleep apnoea, with two times more men than women affected. However, a large number were undiagnosed.[4]

In 2017 in Australia it has been estimated that 39.8% of adults experience some form of inadequate sleep. Those Australians suffering excessive daytime sleepiness (EDS) comprise 19.1% of which 5.8% suffer EDS due to sleep disorders, namely insomnia, restless legs syndrome (RLS) and obstructive sleep apnoea. In total an estimated 7.4 million adults did not regularly achieve the sleep they need. This inadequate sleep was estimated to result in 3,017 deaths in 2016-17.[5]

My mother had severe sleep apnoea but no-one, including her doctors, had diagnosed it. So her condition remained untreated. The debilitating effects of sleep apnoea on her body and mind were also unknown. My mother continually suffered from "bad nerves" or hypertension. She was overweight and snored loudly. She had restless leg syndrome[6] which caused her to involuntarily kick my sleeping father, who ultimately vacated the matrimonial bedroom in search of a good night's sleep.

We all need a good night's sleep. Our brain and other essential bodily organs depend on it. Otherwise we are unable to function on a daily basis as we should. Without quality sleep we struggle through our daily tasks, sometimes involuntarily succumbing to bouts of sleep or "power naps" as they are euphemistically known. The very real danger presented by these attempts by the brain to recapture lost sleep is that they can occur with little warning and at any time. They can also occur whilst performing any task.

This story is a true account of my journey in struggling with and confronting sleep apnoea. Although reference is made to scientific terminology and medical methods of treating sleep deficiency (both conventional and unconventional) it is essentially my personal narrative of how my body and mind reacted to the treatment I received. It recounts my attempt to control this increasingly common breathing disorder.

My sincere hope, and my ultimate reward, from writing this book is to inspire those who suffer from sleep apnoea to take control of their condition. In doing so you will *begin* to know your body and how it functions. You will learn to optimise its functionality and enjoy the benefits which flow from it.

My primary wish is to free those people from their dependence on respiratory aids and to liberate them from their breathless sleep.

1

The many tentacles of the "apnoea octopus"

Sleep apnoea is a pause in breathing during one's sleep pattern. The word *apnoea* (US ***apnea***) medically means a temporary cessation of breathing. The word comes from the Greek word *apnoia* which, in turn, comes from *apnous* meaning breathless.[7] It can last several seconds or several minutes. In extreme cases it deprives the brain, heart and other essential organs of oxygenated air necessary for the proper functioning of our body. As we all know, our body and mind need the restorative effects of a good night's sleep.

I can recall, when I was in my thirties, being roused to partial consciousness by my wife because I wasn't breathing. "Paul – breathe!" she would exhort from some far distant valley. After processing my scrambled thoughts I rolled over and wondered what she was carrying on about. Invariably, I would later discover, I had been snoring whilst lying on my back ... until I stopped breathing. This abrupt halt to my breathing caused my wife some alarm. However, I was largely oblivious to what was happening – let alone *why* it was happening.

In later years I would learn that snoring is the body's attempt to slow down the rate of breathing. If we overbreathe or hyperventilate, we deplete our lungs of the amount of carbon dioxide (CO_2) necessary to regulate our breathing and maintain the pH (or proper acid/alkaline balance) of our blood. The maintaining of the correct pH levels is essential for all living organisms.[8]

I would also learn that the sudden and abrupt interruption to my breathing was a signal from my body that its lungs needed to replenish the amount of CO_2 necessary for efficient breathing.

During my thirties I was overweight. I was married with two young children. My wife worked irregular hours and I was a self-employed lawyer. The recession in 1991 forced me into sole practice. Life was challenging enough without the debilitation caused by sleep apnoea.

Unconsciously I continuously breathed through my mouth both day and night – like an animated goldfish. But just as a goldfish cannot be sustained by breathing through its mouth, neither could I be adequately sustained. Yet I had done so all throughout my childhood and teenage years and right through until my late forties. The only time I can recall it ever bringing itself into focus was when I was a teenager and a friend of mine punched my arm when I ridiculed him for breathing noisily through his nose. We were at a funeral and the noise made by his whistling nose jarred with the otherwise funereal silence expected on the day. Also, his nose breathing annoyed me. It struck me as an unnatural aberration. Didn't everyone breathe through their mouth?

Yet now I would no sooner breathe through my mouth than put my hand in a frypan of hot oil. In the space of nine months – a mere pregnant pause – I managed to overturn a lifetime habit of mouth breathing and convert to exclusively breathing through my nose.

With the benefit of hindsight I can now reflect on how sleep apnoea impacted on my quality of life. As a young man in my thirties I would often fall asleep in front of the television in the evening. I also experienced, on more than one occasion, episodes where my heavy eyelids would stay closed for many seconds whilst night driving. At times my car would swerve off the road forcing me to quickly correct the steering wheel. I would wind down the window and drive with a wall of cold air reviving my sleepy brain. The worrying aspect was that these episodes sometimes occurred when my children were in the car.

In fact, studies have attributed more than 20% of road accidents to driver sleepiness. Sleep apnoea, affecting about 25% of middle-aged men, has been identified as an important cause of driver fatigue. Drivers with sleep apnoea have shown to be at 2–7 times increased risk of motor vehicle accidents compared with drivers who do not have sleep apnoea.[9]

It has been my experience, whilst driving, to momentarily fall asleep without warning. Even though I was consciously aware of being relaxed and tired I had no warning whatsoever of the possibility that my eyes would actually close and that I would succumb to a microsleep.[10] Even after experiencing such an alarming and frightening episode I recall treating it as a total aberration and being utterly convinced that it would not happen again.

In an article in the 2003 *Medical Journal of Australia* on fatal fall-asleep road accidents[11] it is observed that studies have shown that healthy people do not fall asleep without a significant awareness of sleepiness for some time before a "fall-asleep episode". However, it is important to note that this research on awareness of sleepiness was conducted on healthy volunteers, not patients with sleep disorders. Patients with sleep disorders, the article states, may not be aware of impending sleep. It further states that patients with sleep apnoea are often only aware of the severity of their sleepiness *after* treatment of their sleep apnoea.

I can confirm this ignorance. For many years my sleep apnoea was undiagnosed and I struggled on blissfully unaware of how serious a threat my condition posed to me, my passengers and other drivers and their passengers.

There is another less obvious and little known consequence of sleep apnoea insofar as the male's libido and sexual performance is concerned. To my great dismay, horror and shame I suffered from frequent bouts of impotence during my first marriage. I was under 40 years of age! I was completely ignorant of any connection between my sleeping patterns and my impotence. I had blamed it on other elements of the relationship. The psychologist, from whom I sought assistance for my ailing marriage, was also ignorant of any connection between impotence and sleep deprivation.

For sexual arousal and erectile function to be maintained it is essential for your nervous system to be relaxed. The consequence of a breathing disorder such as sleep apnoea is that it is associated with stress responses, including increased heart rate, blood pressure and blood supply to vital organs like the brain and lungs and away from low priority tissues, which include the reproductive organs.[12]

In the context of a Korean study on the link between depression and erectile dysfunction in men with sleep apnoea it was posed that sleep apnoea might impact what is known as the parasympathetic nervous system, which is responsible for many bodily functions including sexual arousal.[13]

During the day I would often yawn or sigh heavily without being aware I was doing it unless someone mentioned it to me. Sometimes I would put my head down at the office during my lunch break and succumb to a 15, 20 or 30 minute sleep. I would not feel refreshed afterwards but would regain some semblance of normal functioning. Often my thoughts would become fuzzy or scrambled. I would put this down to a lack of concentration or just the effects of a bad night's sleep. Having a bad night's sleep would eventually become a normal phenomenon for me, particularly as I struggled to cope with depression following the deaths of my parents over a 16 month period during my mid-40s.

I often suffered headaches and tried to alleviate them by resorting to "sugar fixes" or "popping a pill". I was self-medicating without understanding why my body was behaving in the way it did.

For years my skin, particularly on the soles of my feet and my scalp, was dry and flaky. Skin deposits formed on my legs. I had accepted this as "part of my lot" or just the way I was.

With hindsight I was amazed at how I battled on unquestioningly, enduring my pains and discomfort – almost like a dumb animal. I have known others, including family and friends, who also struggle on and tough it out. Perhaps we do so because we are creatures of habit? In doing so we live by the maxim "ignorance is bliss" and, at the same time, become beasts of burden.

My lack of sleep did not inspire me to exercise as my energy levels were low. So instead of shedding weight I actually gained weight. However, my 15-year-old son was keen on going to the gym. So I dragged my weary body and accompanied him by going through the motions on the various exercise machines. I would much rather have slept. But I was also torn between wanting and not wanting to sleep. I was not enjoying good quality sleep and I felt that I might jeopardise my prospects of sleeping later that night.

So I was caught on a vicious merry-go-round which made me feel dizzy and light-headed. These sensations also made me short-tempered and irritable. I was often subjected to panic or anxiety attacks.

On other occasions, without warning, I would feel a chill of fear from within my soul. It sometimes took hold of me and catapulted me into absolute despair. The depression didn't fix itself on anything in particular, but latched on to everything that was going on in my life. Whilst I was feeling this way I would paint black any aspect of my life which I focussed on.

A study of 18,000 Europeans conducted in 2003 and appearing in the Journal of Clinical Psychiatry is the first study to show a link between depression and sleep apnoea. People with depression are five times more likely to have a breathing-related sleep disorder than non-depressed people.[14]

Lack of regular quality sleep had the effect of exacerbating everyday stresses of my life. It magnified the losses I had experienced. Those losses began with the fragmentation of my family following the breakdown of my marriage in 1999. I found it very difficult to cope with being apart from my two young children (then aged 3 years and 6 years). I would often refer to them as "my left and right hands".

For the next eight years or so I struggled with tending to the needs of sick parents, especially my mother. They would eventually leave this world less than 16 months apart from one another during 2006–2007. Their absence left a huge hole in my life both separately and jointly. I was fortunate to be in a loving relationship at the time each of my parents passed on. Those relationships, at the time, provided significant support for me.

My life was full. I was both managing and conducting my legal practice, bringing up children, maintaining a relationship and looking after sick parents. Consequently it was easy to push aside those health issues which didn't immediately confront me or prevent me from functioning at all. Sleep apnoea can be surreptitious as, depending on its severity, it may only slightly or moderately retard your functionality. I could largely ignore its deleterious effects as I was still able to function or "cope" even though, with hindsight, not to my optimum level.

I believe that during the three years from 2005 to 2008 sleep apnoea began to assert itself more prominently. During a time of emotional and physical stress bad breathing habits began to exact a toll. At one point the combination of antidepressants and sleeping pills with sleep apnoea meant that I was only having two to four hours of sleep each night. It was a rollercoaster of nightmares which I can barely recall thanks to the passage of time and the cloudiness of an unclear memory.

Surprisingly, it wasn't until 17 July 2008 when my treating doctor first diagnosed the possibility of a breathing disorder, in particular sleep apnoea. I had attended a Council of Adult Education (CAE) seminar on breathing habits a few weeks earlier. Consequently I realised that I needed to undertake a sleep study.

It also led me to make further enquiries which would ultimately prove to be life-changing.

2

Facing and conquering my worst fears

My interest was aroused in about the middle of 2008 when I read that the CAE was holding a free introductory seminar on Buteyko (*Bew-tay-ko*) breathing. It was claimed that this method of breathing successfully treated people who suffered from asthma and sleep apnoea. Although I knew very little about sleep apnoea, I knew my sleep was interrupted. I suspected that I may have been ceasing to breathe at night. I knew my breathing was irregular. I wanted to know more.

I sat towards the rear of the room. Mr Paul O'Connell, Director of Buteyko Health & Breathing, gave the presentation. He was a strong and fit looking man. He spoke well. He was a former asthma sufferer who had been cured by Buteyko breathing. He held a Bachelor of Science, Diploma of Education and a Master of Business Administration.

We were told that our medication should not be thrown away and that our doctors should be consulted regarding any treatment that we contemplated undertaking. He spoke of the history of Buteyko breathing and its pioneer Dr Konstantin Buteyko, a

Ukranian who developed the reduced breathing method in the 1950s. As a medical student he studied the breathing habits of dying patients. He noticed that they tended to overbreathe. He pondered whether disease led to overbreathing or whether overbreathing led to disease. His experiments confirmed the latter.[15]

I found it a little difficult to relate to the discussion about breathing methods for asthma sufferers as I had never suffered with asthma. But when the discussion turned to sleep apnoea he had my full attention. We were told that by sleeping on our side (preferably on the left) with our head raised on two pillows we would experience an almost immediate improvement within the first 24 hours.

Lying on the left side was preferred over the right side as everything is more compact and fits better. The heart is stabilised and the left lung is smaller than the right lung as it shares space with the heart.

I understood that elevating your head whilst sleeping on your side helps keep the airway open thereby assisting to prevent your airways narrowing or the walls of your throat from collapsing.

Mr O'Connell explained how the Buteyko breathing course involved reducing the depth of breathing and retraining our brain to be accustomed to higher levels of CO_2. This was achieved partly by breathing exclusively through our nose. We were told that the nose is designed to act as a filter and humidifier of air before it enters our lungs. We learnt of the effects of breathing too much air through our mouths. Snoring, it was suggested, was one of the body's ways of attempting to slow down our rate and restrict the volume of air we breathed as it tried to hold on to $CO_{2.}$

Confirmation of the existence of sleep apnoea could only be scientifically confirmed by undergoing a sleep study. This is a process where trained sleep technologists monitor your sleeping patterns by means of electrode wiring attached to your head, chest, legs and fingertips. I will describe this fascinating experience later in this book.

We were told of the following symptoms of sleep apnoea:

- Apnoeas from 10 to 180 seconds
- Snoring
- Excessive movement and restlessness
- Breathing through the mouth
- Dry mouth and throat during the night and on waking in the morning
- Thirst overnight and on waking
- Waking feeling unrefreshed
- Daytime fatigue
- Poor daytime concentration
- Falling asleep in meetings, watching TV or whilst driving
- Breathlessness at rest or when exercising

As I listened to these symptoms being recounted I heard a checklist of my symptoms. I was shocked. I had been unaware of what was causing my fatigue and relatively poor health. I was totally out of tune with my bodily functions, including its shortcomings and capabilities.

We were told that the course would teach reduced breathing by means of controlling the amount we breathed and slowing down the rate of our breathing. Shallow breathing or breathing with our upper chest was to be avoided as it led to hyperventilation. We were directed to our diaphragm and the benefits of deep and slow breathing.

It was suggested that, depending on a person's symptoms, health conditions and medications, we could assist nocturnal nasal breathing by taping our mouth closed at night. This should occur until our brain was "trained" to only allow breathing through our nose.

I was sceptical of the concept of taping my mouth at night. I had always breathed through my mouth. I could barely conceive of how I could psychologically deal with such a radically different approach. It was like being told you had to convert from writing with your left or right hand to writing with the opposite hand.

Others within the wider class appeared to share my scepticism. Some members of the class pondered whether we could learn much more than what we had been told already. Mr O'Connell's response was that the course would provide important instruction and guidance on the reduced breathing techniques. The key was to change the way we breathed in the daytime so that we change the way we breathed at night.

That night I resolved to put into practice some of the advice I had been given. I slept on my side with my head raised on two pillows. I could not entertain the thought of mouth taping. In fact I had rejected the concept completely. The next morning I noticed a definite improvement on wakening. The fog in my head had cleared noticeably. It was remarkable to experience the difference. However, with hindsight, the difference must have been minimal. I suspect that the impact of those small changes in my sleep behaviour indicated how serious my sleep apnoea had become.

On some nights I would only sleep with one pillow. My reasons for doing so were varied. Sometimes I wanted to challenge the condition and I chose to sleep "normally" with one pillow as I had always done. On other occasions I cynically believed that two pillows were not making the difference I had hoped for and I would often "reality test" it by only sleeping with one pillow. Sometimes I would feel rested the next morning but often I did not feel rested and I was often plagued by a headache – a band of pain that formed a ring around my head. The irony, I would later appreciate, was that although I used to sleep in a *normal* fashion the quality of my sleep was not "normal".

Knowing what I know now it is difficult for me to understand how my attitude towards sleep apnoea could be so ambivalent. But then I consider that at that stage I continued to mouth breathe and my body had sadly become accustomed to my tissues receiving a lower saturation of oxygen.

Consistent with being a creature of habit, I stumbled along relying on what I had learnt at the CAE seminar and hoping things would improve. They didn't. My weight had increased and the quality of my sleep decreased. So I saw my treating doctor and requested a sleep study so that I could have my condition confirmed.

I saw a sleep specialist who arranged for me to undergo a sleep study at Vaucluse Hospital in Brunswick on 14 October 2008. I arrived shortly prior to 7.00 pm and was shown my room. It was sparsely furnished with a small bathroom en suite. I was weighed, measured and had my pulse/heartbeat recorded.

As the time to sleep drew closer I was fitted with wires and other recording devices designed to record my pulse, rate and volume of breathing, bodily movements and brain activity during my nocturnal odyssey. I asked for a mirror after the hook up was completed. I looked like “an extra” for a Frankenstein horror film! I snapped an iPhone photo of myself and sent it to bemused friends and family.

I found it difficult to sleep as my movements were restricted by the wiring. When morning arrived I felt that I had slept poorly.

My vital statistics were:

Weight:	95 kg
Body Mass Index (BMI):	31 kg/m^2 (obese)

My results as detailed in the attached Diagnostic Sleep Study Report[16] were:

Sleep period:	432 min
Time awake:	44 min
Rapid Eye Movement (REM) sleep:	82.5 min
Sleep efficiency:	81.8%
SaO_2 % min average	92.0%
SaO_2 lowest	83.0%

An apnoea of 29 seconds was recorded.

My eyes scanned the above details with little acknowledgment or understanding. My blood froze when my gaze fell on the following line:

Diagnosis/Treatment

"Severe REM related SAHS. Recommended CPAP."[17]

Well, I was having none of that! The thought of being chained to a machine which manufactured Continuous Positive Air Pressure filled me with dread. I recoiled at the image of being chained to the machine. I appreciate that for some people CPAP may well present itself as a lifeline. But I was simply unable at that stage to psychologically entertain the concept. Following a debrief with my sleep specialist and a detailed discussion of the options available to me, I chose a mandibular splint or mouthguard. This implement was intended to set my jaw forward and prevent my tongue blocking my airways.

I made an appointment to see a prosthodontist. He needed to make a replica mould of my upper and lower levels of teeth and gums. There were metal clips or hooks which connected the upper and lower sections of the mouthguards.

As I lay in bed with my jaw slightly jutting forward I felt uncomfortably ridiculous. Talk about leading with my chin! But with time I began to grow accustomed to it. The accumulation of saliva during the night and the need to regularly clean it annoyed me but it appeared to assist me to breathe more easily through my mouth. I say that it appeared to assist as I was also sleeping on my side in an elevated position. I persisted with the mouthguard for about 6 months until the benefits seemed to plateau.

My weight was still about 95kg.

In September 2009 I resolved to improve my fitness. I went on a holiday in Fiji with my partner Deb (now my wife) and our children. Through my participation in water sports, including diving, I managed to shed some weight. And I slept better. Perhaps I was getting on top of the dreaded apnoea?

My 50th birthday was approaching. Apart from planning to throw a party for family and friends I wanted to do something I had always dreamed of doing, that is, dive the Maldives. From 1986 onwards I had sampled (on many occasions) the wonders of the Great Barrier Reef. I often imagined diving in two other famous locations. Those dive sites were the Red Sea at Sharm El Sheikh in Egypt or somewhere amongst the myriad atolls of the Maldives.

So I discussed my dream with Deb who was very keen as she had obtained her open water dive ticket only a few months ago. On the night after the party we flew to Male via Singapore. During our stay we dived regularly. The regular exercise and healthy eating in a stress free environment enabled me to breathe in a more controlled manner. I did not use the mouthguard and I continued to mouth breathe. With hindsight I now understand that my carbon dioxide levels had increased through the exercise and that such an increase assisted in the efficient regulation of my breathing. The conditions within my lungs and distribution of oxygen to my tissues must have been about right as indicated by my general sense of well-being and presence of mind. I will delve further into the science of the matter in Chapter 7.

But following our return to Melbourne and resumption of our busy working lives, the quality of my sleep began to deteriorate.

I gained weight and my headaches returned. Each morning I drove my daughter across town to her school and then drove on to my city office. This two hour trip began to exact a toll on me. I developed a persistent ache in my right hip from sitting in the car for so long. I continued to breathe deeply and exhale too much CO_2. At times I would almost fall asleep at the wheel whilst waiting at a set of traffic lights. I momentarily closed my eyes and would often ask my daughter to tell me when the lights changed from red to green!

When I finally arrived at work it felt as though half the day was over. On some occasions I returned home to sleep or to try to sleep, perhaps to breathe ... that was the challenge! But I couldn't rid myself of the persistent headache. Quality sleep eluded me. The effects of poor quality sleep were wearing me down. I simply had to seek some help.

It was in March of 2010 when I sought help from a sleep disorder clinic. My treating doctor was empathetic. He looked as though he may have had difficulty sleeping too. But I recall that he had young children and those of us who have gone through those years know what a sleep deprived time of your life it can be! I thought I had moved on from that experience when my children grew into independent sleeping beings. I told my doctor of my brief experience with Buteyko breathing and my desire to delve into it further. He appeared genuinely open to me doing so as he had some knowledge of Buteyko breathing but he advised I needed some fairly urgent relief. I was again prescribed a CPAP machine.

My resistance was very low and I resigned myself to submitting to CPAP on a trial basis. I decided to rent a CPAP unit and hoped that it would return some lost sleep to me, provide my frazzled brain with some clarity and bestow some tranquillity upon my tortured soul.

Continuous Positive Air Pressure is what CPAP promises. It consists of a mask connected to a flexible silicone hose that plugs into a breathing apparatus which emits pressurised and humidified air into an open mouth or nostrils. There is a setting on the machine which regulates the rhythm and frequency of the air inflow. It also has the ability to automatically adjust to your particular breathing pattern.

I was ready to try it. In fact, I was so punch-drunk through a lack of sleep and the resultant lack of oxygen to my brain that I would have tried anything. I quickly made an appointment with a CPAP supplier and selected a satin unit and mask. The machine was priced to sell at $2,448. I elected to rent it for one month. I had to purchase the mask. I also needed to purchase a chin strap to prevent my mouth from opening. The chin strap ran along the sides of my head and was secured to the crown of my head by velcro attachments. It looked like a head brace used when one was recovering from a severe head injury – which was how I sometimes felt courtesy of my persistent headaches.

I left with the packages tucked under my arm after paying $375 for the privilege. I later discovered that although it was possible to claim under my private health insurance for the cost of purchasing a CPAP unit, there was no possibility of claiming for the purchase of the mask or the rental cost of the unit.

The confrontation

I sat on the edge of my bed. With shoulders slumped I faced the CPAP machine. It was satin chrome. It looked sleek. It would be quite at home in any modern setting as an elegant device imitating a wireless sound system. But I saw the chameleon for what it was. CPAP was a device not for music but a device for forced breathing.

A current of fear radiated across my stomach as I placed the silicone mask over my nose and mouth and fitted the straps around the base of my skull. I fitted the head brace to secure my jaw. I lay on my back with my head on two pillows. Turning my body on my left side and facing the machine, I pushed the switch. It activated a soft blue light.

I waited with anticipation.

Soon a warm current of air flowed into my nostrils. "Don't fight it," I commanded. "Go with it. Let it come."

But my mind soon rebelled against the contrived rhythm it imposed on me. "What if I don't want or need to breathe out when I am supposed to? What if my lungs are not ready to inhale the next forced airflow? What if I don't want to dance to the tune of the sleepmaster?"

I knew that it was all a matter of conquering my anxiety and somehow relaxing. But I had great difficulty in quelling my indignation. Why was I so indignant? I will tell you why. When you have been breathing independently for almost 50 years it is very difficult to accept that you must now be assisted in doing what all healthy able-bodied people take for granted every sleeping minute of their lives. When you are fatigued and in need of sleep, you do not want to be preoccupied with the prospect of not breathing correctly. Life is difficult enough at times and sleep, for most of us, should afford an escape from the stress and anxiety of everyday life. Sleep is the essential elixir we need to carry on with our lives the next day.

At times the airflow malfunctioned and air gushed as if escaping from a punctured tyre. Frustrated and feeling helpless, I sat up and reset the machine. A sharp headache compounded my growing sense of hopelessness. I worried about my confidence being undermined insofar as breathing without the machine was concerned. I worried about my lungs not being able to do at night what they have been doing since my first newborn slap!

Hours passed as I struggled throughout the night with the machine and all it represented. I felt beholden to the machine. I was chained to it. The machine was both my saviour and my millstone. I was a slave to it yet I had also enslaved it to me.

Did I control it or did it control me? Did I care anymore? Overwhelmed by exhaustion and a lack of quality REM sleep I yielded to its satin charms. I succumbed to the breathing pattern forced upon me. And I finally slept. It was a profoundly dreamless sleep.

On awakening I felt refreshed. My spirits were inflated and I felt insulated from fatigue. The machine had heralded a new beginning for me. It gave me a chance to function normally throughout the day. I felt a renewed sense of energy as my blood had been oxygenated.

Over the next few weeks I would take my CPAP with me on the plane when I visited my partner interstate. During my honeymoon with CPAP I was actually quite proud of it. It had given me a new lease of life. I had found reserves of energy where previously there had been none. There were many positives both physically and psychologically. However, over time, I could sense a growing dependency on CPAP – and this alarmed me.

There were also other negatives associated with CPAP. The hosing made it restrictive when moving from side to side and it appeared to be designed for sleeping whilst lying on your back. I am yet to see a photograph of a CPAP user lying on his or her side. It was a position which was opposite to the position I later learnt was the best position to adopt when sleeping.

The mask, whilst initially comfortable, became irritable to my skin and the enclosed area around my mouth and nose was claustrophobically humid and wet with exhaled condensation. Most worrying was the contrived rhythm which CPAP imposed on my breathing. At times I felt like I was drowning. It did not appear to provide the benefits of a variation in the pace or rhythm of breathing which often occurs during a REM induced dream. I found the apparently forced unregulated pattern of breathing alienating. I felt like an invalid who had no choice but to submit to the will of the CPAP – and be eternally grateful for it.

My partner was completely supportive of the benefits CPAP brought me. Even though she also appreciated being able to enjoy an improved quality of sleep her support was important to me. But I was also aware of CPAP not exactly having any aphrodisiacal visual qualities and, it must be remembered, our relationship was relatively new. Over time the novelty began to wear thin.

I had a growing belief that my relationship with CPAP could be a fleeting acquaintance. It was a bit like experiencing a whirlwind romance that makes one giddy with the excitement of new love but which ultimately proves to be unsatisfying.

I knew that I had to find another way of fighting the dreaded apnoea.

Following my attendance at the one-off CAE introductory seminar on the Buteyko method of breathing, I resolved to do the course before fully committing to CPAP. When I looked at the course timetable I was dismayed to see that the five day course ran only during the week! I simply could not consider taking off the whole week from my practice.

It seemed that the course was targeted to mature age persons who could afford the time to attend. Perhaps this age group was the medical demographic? I had thought that sleep apnoea only afflicted people of my age and older. However, I was aware that asthma affected people of all ages. I later discovered that sleep apnoea can afflict young people in their twenties! I also learnt that although reducing weight can reduce the effects of sleep apnoea, it can afflict people who do not appear to be overweight.

So I researched the internet and found a Buteyko practitioner who could see me not only after hours during the week but also on the weekend! She was also prepared to spread out the sessions over many weeks or however long it took due to my personal circumstances. This was very convenient and motivated me to start the course as soon as possible. I was impressed with her qualifications in biochemistry and her knowledge of the respiratory system and its effects on the body.

My first appointment was at a home office in an inner-city suburb of Melbourne. She explained to me that the course involved retraining or reconditioning my brain to breathe exclusively nasally and at a slower rate. I was told that many modern ailments can be attributed to overbreathing. One of the aims of the course was to prevent CO_2 levels from being depleted and to maintain reasonably high levels so as to allow oxygen to be released to the tissues. I later learnt about the relationship between the body's production of CO_2 and its release of oxygen by haemoglobin.[18]

I vividly recall being asked to sit with spine erect so as to allow as much space between my lungs and diaphragm. My shoulders were to be comfortably squared or "soft". I was advised to look ahead or upwards rather than downwards as the latter action activated the brain's fight/flight impulse with a consequential increase in pulse rate. If my pulse rate quickened it would be more difficult to slow my rate of breathing.

I was then invited to inhale normally through my nose and then let out a little air. Then, whilst facing straight ahead, I was asked to hold my breath by gently pinching my nose. How long could I comfortably hold my breath? I exhaled after holding my breath for 12 seconds as recorded by the stopwatch. I needed to breathe and could not have continued without feeling distinctly uncomfortable. My pulse was recorded as 72 beats per minute.

The process I had just undergone was known as a Control Pause (CP). Its purpose was to provide a reasonably accurate reading of the amount of CO_2 in my lungs. I was told that it was important to inhale at the moment I felt the need to inhale. The goal was not to hold my breath for as long as physically possible. In fact, the first breath at the end of the CP should not be deeper than the breath I took before I commenced the CP.

She told me that my goal would be to hold my breath for about 50 seconds. I smiled derisively and shook my head. I privately thought it an impossible goal. But at the same time I had confidence in my mentor and now I saw a glimmer of hope at the prospect of being able to breathe independently.

I made an appointment to return my CPAP.

I was now excited and inspired.

But I couldn't quite let go of my CPAP yet. Despite my misgivings it had improved my sleep. Although I found it burdensome in the short time we had known one another it had become my security blanket. If given the choice between a night of broken sleep with its attendant anxieties and the forced cushion of warm air supplied by CPAP, I would always choose CPAP.

To endure a night without quality sleep is to travel a mentally tormented landscape filled with potholes of fear and ravines of self-doubt.

The use of sleep deprivation as a means of torture for prisoners of war is outlawed by the Geneva Convention.[19]

So why would anyone voluntarily endure it?

The breathing process first involved checking and recording my pulse rate. Then I would inhale with my back straight and eyes focussed ahead. After having exhaled about a third or slightly less of the air in my lungs I held my breath. I found that it was more secure to pinch my nose with my thumb and forefinger as the need to breathe can be insistent. The urge to breathe can manifest in many sneaky ways if the usual air supply to the lungs is deprived. For example, the body may manufacture a yawn so as to force the mouth open.

As previously stated, I was to look ahead rather than down. Otherwise it could communicate fear to the mind and stimulate the instinct of flight. If that happened my pulse was likely to quicken and consequently produce a faster rate of breathing. I was told to focus on the goal of slowing down the rate of breathing and, in doing so, slow down the need to breathe as frequently.

A slower rate of breathing creates a more conducive environment for the lungs to retain CO_2 rather than "blowing it out".

At about 11.30 pm on 6 April 2010 in the quiet and comfort of my bedroom I began my first reduced breathing session (RBS). My resting pulse rate was recorded as 72 beats per minute. My first held breath or CP was for an unspectacular 10 seconds.

Next I began my 5 minute RBS. It was difficult. I felt my head pound as I reduced the amount of air I would let into my lungs. That air which was granted access made it all the way to my diaphragm. No shallow breathing was permitted. Each breath was to be controlled and slow. I instinctively understood that much benefit was to be gained from exhaling very slowly.

I later learnt that the slow exhale of air enabled contact to be made with the alveoli (those little air sacs or cells which hold air in the lungs) which would allow the vital exchange of oxygen and carbon dioxide to occur.

So now I had embarked on a course which would forever change my breathing habits. My mind silently but indignantly screamed at my rationing of air. My heart rate quickened and my body temperature rose as I gradually slowed my breathing and decreased the amount of air I inhaled and exhaled.

I struggled against the urge to take a deeper breath than I permitted myself to take. My next control pause was 14 seconds. This was a small and incremental increase but I had to keep reminding myself that this was not a competition. At the first sign of real discomfort I was to exhale and record the time captured by the stopwatch. I would later learn of the psychological games I could play against my mind's insistence that I breathe. I found that it was important to be relaxed and not allow panic or anxiety to dominate.

After each RBS I was to rest for a minute and breathe quietly and slowly. My next RBS resulted in a control pause of 17 seconds. Gradually I began to feel a rhythm to my breathing. The excitement built within me as I began to consciously feel a change. It was not only metaphysical but also physical. I was intrigued by the calmness I felt. Although it was almost midnight

I began one last RBS. Every now and then I could feel a nerve sensation trickle across my temple – like a current of electricity. I chose not to analyse it too much and went with whatever was happening. My next control pause was 12 seconds.

I was feeling tired. I recorded my pulse and discovered that it had slowed to 60 beats per minute. I was feeling tired but relaxed. I was ready for sleep – but not without CPAP.

No not yet. I was unable to quite let it go just yet. I resolved to tape my mouth at the same time as using CPAP. By doing so I would prevent my mouth from opening and I wouldn't have to wear the "ninja chin strap". I applied some medical tape horizontally across my mouth and pulled the CPAP mask on. I shook my head derisively at the thought of what I must look like.

That night I woke only once and slept soundly for 5.5 hours. The next morning I commenced another session of reduced breathing. My CP ranged from 14 to 19 seconds and my pulse ranged from 64 to 68 beats per minute. I was beginning to feel in control of my life and was looking forward to my day.

How long had it been since I last looked forward to my day?

I had been advised not to eat less than two hours before commencing an RBS. Otherwise it would be more difficult to reduce my breathing and to attain increased carbon levels as the digestive process increases the pulse rate. The quickened pulse would, in turn, make it more difficult to slow my rate of breathing.

I experienced precisely this difficulty that night as I began my session at 9.30 pm, less than 2 hours after my evening meal. My heart pounded as I struggled to control my breathing. My CP ranged from 12 to 16. I felt that I had gone backwards in my progress. Somewhat disheartened I retired to bed with mouth taped and face masked. My sleep was interrupted by CPAP that night. At 2.30 am I could stand the mask no longer. It came off. It would be the last time I ever used CPAP. My eight week relationship with CPAP had come to an end. My broken sleep was recorded as 4 hours plus 3.5 hours after removing the mask.

The next day I had three separate reduced breathing sessions with my CP peaking at 17, 26 and 22 in each of the sessions. Within those sessions I noticed that my CP was becoming more consistent. My pulse was lowered by the end of each session by usually 4 beats per minute and on one occasion 8 beats per minute. This indicated that my rate of breathing had slowed. My last session was at 4.30 pm as it was Deb's birthday and we were going out to dinner that night.

Despite having a few glasses of wine (which increases the rate of breathing) I slept for 6.5 hours and felt rested and energised the next day.

My fourth day saw my CP reach maximum times of 17, 25 and 26. The sessions could be peaceful and relaxing or they could be just hard work. I soon learnt that the secret was to patiently persist in continuing to maintain the reduced amount of each intake breath despite an increasing urge to open my mouth and take a huge gulp of air.

As my heartbeat quickened and my temperature rose I struggled to keep my breathing at a slow, steady and even flow. I slept 7 and 8 hours over the next two nights.

Over the next five days I continued diligently with my reduced breathing at least twice but mostly three times each day. My average CP steadily rose over that period from 19.1 seconds to

29.5 seconds. My maximum CP was 39 seconds and my average sleep each night was 5.8 hours.

A quiet revolution had begun within my physiology. My internal processes appeared to be functioning at a far greater optimum level. My senses were on alert and my body moved with greater fluidity. Even the dull ache that emanated from my right hip whilst I drove my daughter to school and then drove to my city office began to dissipate.

My energy levels increased and I was amazed at comparatively how little sleep I needed each night to function throughout the day. When I woke the next morning I felt the benefits of a deep sleep. I remembered nothing of my dreams – yet my calm and restful mood told me that I had enjoyed good REM sleep.

Throughout my day I felt sustained by what I now know to be my increased carbon dioxide levels. I constantly marvelled at being "headache free" (and still do). I simply did not have any headaches. Occasionally I would feel fatigued and experience a brief dull ache – like a reminder of a headache. I would then slow down my breathing and inhale a reduced amount of air. It would not take long before I felt refreshed and centred.

I later realised that these "headache reminders" invariably occurred at a time when I was unconsciously breathing too heavily or quickly.

Almost three weeks into my breathing regime my average CP climbed into the 30s. It was during this time that my Buteyko practitioner introduced Extended Pauses (EP) as part of my exercises. An EP, as the name indicates, involves my CP being extended. I was to extend my CP by 8 seconds after the first sign of discomfort was felt.

My breathing regime involved 4 x 5 minute sessions of reduced breathing, with a CP recorded at the end of each session. My EP was to occur after my third 5 minute session. In the early evening of 24 April 2010 I recorded an amazing 52 seconds on an EP. My eyes were bulging as they willed the stopwatch to reach the 8th second immediately. My heart was racing and my diaphragm was palpitating but I felt elated! I recorded my pulse as having increased from 68 to 76 beats per minute at the end of the session (which included two more 5 minute sessions that yielded CPs of 39 and 29). With hindsight my pulse should have *decreased* at the end of the session if I was breathing proficiently but I sensed that the EP was nevertheless spiking my carbon dioxide levels.

Whilst undergoing these reduced breathing sessions I still sometimes unconsciously inhaled through my mouth. When it happened I felt alarmed. Later I would feel guilty if a breath surreptitiously entered my mouth. I soon came to accept that it was all part of my brain being rewired and changing a lifetime habit of mouth breathing. At the beginning I actually felt a deep sense of loss at the prospect of never again breathing through my mouth. I enjoyed mouth breathing.

There were times when I felt that my nostrils were not wide enough to allow sufficient air to reach my lungs. The greater the volume of air I quickly sought to inhale, the narrower my nostrils became. This was an alarming experience which was exacerbated when feeling anxious or stressed.

When I scuba dived I necessarily breathed through my mouth. The thought of mouth breathing would fill me with a sense of trepidation as I feared that I would begin to let loose the accumulated levels of carbon dioxide I had stored. But I had no choice as nasal breathing was not an option when scuba diving. However, when muscles move they produce lots of carbon dioxide, which helps to maintain the normal volume that is required to stay healthy.[20] This fact was confirmed by my Buteyko practitioner who told me that kicking one's legs, in particular, was an excellent way of building up carbon dioxide levels.

It was my experience that after diving I was easily able to revert to nasal breathing without feeling that I had regressed in any way whatsoever. When I scuba dive now I secretly enjoy the sensation of mouth breathing. It is also of some comfort that this activity does not have a regressive effect on my CO_2 levels.

During the fifth week on 9 May 2010 I held my breath for 60 seconds for the first time. It was an EP but it paved the way for me to achieve a CP of 60 seconds on 14 and 15 May 2010 at the end of my RB sessions.

Three days later on 18 May 2010 at the end of my usual 4 x 5 minute RB sessions I achieved an astonishing CP of 70 seconds. I recall a distinct feeling of oneness between mind and body. You may call it a Zen moment but I felt wonderfully alive and it seemed that all my senses were dilated and receptive to an inner energy.

I was not gasping for breath – quite the contrary – I felt surrounded by air but felt little need to inhale or exhale other than in an almost imperceptible way. During the nine week period up until 5 June 2010 I diligently recorded my RB sessions making reference to times, pulse rates, CP, EP, hours slept and number of wakes during the night.[21] I also recorded the type of food I had eaten the night before.

3

A now and Zen experience – bodily changes and the benefits of quality sleep

My body reacted to the new regime I had imposed on it. It began to react in many different ways. The first change I noticed was a nauseous feeling and stomach-ache which accompanied the insistence of my bowels to empty themselves on a more regular basis. During the first few days I noticed that my bowel motions were very loose – almost like diarrhoea. The odour was foul. It was as though I was experiencing some sort of detoxification.

Thankfully this experience was only temporary. After six or so bowel movements I noticed an absence of the overtly offensive odour I had detected previously. I appreciate that diet may be a contributing factor in faecal odour. But what I was experiencing to a greater degree, as my carbon levels increased, was an odour which I can only describe as oxygenated. It was as though my faeces were suffused with fresh air!

As my CO_2 levels increased, the laxative effect on my bowels was unmistakeable.

I am reminded of how, as a 10-year-old, I would sit on the toilet in total darkness. I would imagine that I was alone in the house. The noises I would hear of my family moving about the house became the sounds of robbers moving from room to room. My heart pounded as I prayed that I would not be found. It was not long before my bowels loosened. But, in hindsight, it was not so much my wild imagination which opened the stool gates. I now recall that I would significantly reduce my breathing to half or quarter breaths. I would hold my breath as I heard the robber outside the toilet door. I even saw in my mind's eye the door knob being slowly turned. At this point my first unconscious control pause resulted in the emptying of my bowels.

My heartbeat also increased during my RB sessions and my body generated a significant amount of energy. After a few weeks of regular and persistent reduced breathing accompanied by exclusive nasal breathing day and night, I began to lose weight. I eventually lost almost 10 kilograms and was forced to replace much of my wardrobe.

Nasally speaking

At times my nose would become blocked either with mucus or by an apparent total closure of my left nostril. It would happen during the day and night. If my breathing was quicker than it should be, I found it both distressing and frustrating when it occurred at night as I slept with my mouth taped!

Buteyko teaches one how to unblock one's nose by the following method:

1. Inhale and exhale through the nose keeping your mouth closed.

2. Gently pinch your nose with your thumb and forefinger.

3. Whilst holding your nose and breath, nod or shake your head until you can hold your breath no longer and need to inhale.

I have found this method very effective in unblocking my nostrils – particularly my left nostril which was diagnosed as housing a deviated nasal septum.[22] In other words, the centre wall of my nose was crooked and it was restricting airflow through my nasal passageways. When it was blocked I noticed that the more forcefully I exhaled or inhaled air through my nose the more impenetrable my nasal blockage seemed!

As I made my transition from mouth breather to nose breather my snoring changed from my throat to my nose. It occurred because of my deviated septum and my rapid breathing. My snoring was caused by the *Venturi effect* which results in an increased turbulence or vibration in direct proportion to a stronger airflow through a narrow passage.[23] A helpful analogy is the experience of trying to keep a steady gait while walking along a narrow lane between two high rise buildings when a strong wind is howling through the narrow passage.

I later learnt that when I significantly slowed down the rate of my breathing my nostrils dilated. I already knew that slowing down the rate of breathing was conducive to increasing the blood levels of CO_2. What was the connection between increased CO_2 and dilation of nostrils?

I explore this question in Chapter 7. I was amazed to learn that whilst sleeping on one side my nostril closest to the pillow would block, thereby giving it a rest while the other nostril "breathed".

I was informed by my ear, nose and throat (ENT) specialist about the "nasal cycle" which is a natural phenomenon that occurs in the nose during sleep. The brain preferentially sends more blood to one side of the nose at any one time. The blood flow is shunted from one side to the other after intervals of several hours. This gives the sleeper the feeling that the nasal blockage fluctuates from side to side. I have certainly experienced this sensation after I turn in bed from one side to the other.

I was also told by my ENT specialist that the exact reason for blockage of alternate nostrils is unclear.

However, there is a scientific theory that it is only necessary to breathe through one nostril to meet the demands of metabolism during sleep and this is normally achieved by lying on one side. As the nostril closest to the pillow fills with fluid, the other is breathed through. When the working nostril becomes tired it causes the person to roll over and the reverse happens.[24]

The repetition of this cycle ensures a sound sleep as backache, cramp, numbness and circulatory problems can occur if the sleeper stays in the one position for too long.[25]

As an interesting aside, I have found that by reducing my breathing as I lie on my left side, my blocked nostril gradually opens up as my CO_2 levels rise. This allows me to breathe through both nostrils. I do not know for how long I can breathe through both nostrils whilst I am asleep but I suspect that my nostril closest to the pillow would fill up again as my nervous system shuts down during sleep.

Apart from having to deal with a partially blocked nose I also had to somehow accommodate a frequently bleeding nostril. I would often notice a bloody discharge following the unblocking of my nostrils. I was informed by my Buteyko practitioner that a membrane had ruptured, most likely having been caused through continuous, and at times rapid, nasal breathing.

Previously my nose had been practically dormant. As a chronic mouth breather my nose had only been used to smell odours. Now it was being used constantly – and during my transitioning phase air was being inhaled at a quicker rate than was necessary. Considering the sensitivity of my nasal membranes it was perhaps unsurprising that a bloody discharge occurred. I am pleased to report that this eventually ceased.

During the process of adjustment or detoxification as I practised reduced breathing, I was often inconvenienced by a post-nasal drip or runny nose.

Breathing through one's nose seems a simple enough concept to follow. But when you can only breathe through your nose the proposition initially became very difficult. I discovered that my body soon deployed some sneaky tactics in an attempt to coax me back into mouth breathing. For example, I would find myself yawning or coughing which would make it easier for me to mouth breathe.

Habits acquired during a lifetime of mouth breathing needed to change if I was going to have any chance of regulating my pattern of breathing and preventing excessive amounts of CO_2 being blown out. Through patient observation I became aware that I often drew in a sharp breath if the shower water was too cold or too hot. If I spoke too quickly, particularly when becoming excited, or if I laughed, I noticed that I took multiple mouthfuls of air.

If I ate too quickly, whether through hunger or impatience, I gulped mouthfuls of air as I swallowed mouthfuls of food. In doing so I contributed to my stomach bloating as it struggled to deal with digestive gasses combined with the additional intake of air.

If I was very tired or stressed I was tempted to sigh (as I had done for many years) and, in doing so, I would allow a significant volume of CO_2 to escape.

I gradually discovered that if I changed the pace of my lifestyle to suit the pace of my breathing I could significantly improve my chances of overcoming these sneaky bodily tactics. I had always been a rather impatient young man, particularly in my 30s and 40s, always trying to cram too many tasks and activities into my working and social life. Irrationally it was almost as though I tried to defy time. Often I left very little margin for error in meeting deadlines. My life was almost manic as I juggled the demands of a busy legal practice with family life, family illness and relationships. The stress that my lifestyle created resulted in a perceived need for an urgent supply of air. There was simply no time to slowly breathe through my nose – even if I had thought of it!

Many of us, I suggest, need to pause and “smell the roses”. We do not have to be enslaved by an artificial sense of urgency. Do we really have to return that message now? Do we really have to take that call or read and respond to that email we just received less than 30 seconds ago?

Unfortunately the pace of technology feeds our desire to achieve more. We believe that we can achieve more as we complete the many tasks that our computer software programs allow us to achieve. But working at such a relentless pace can have its disadvantages. An overstimulated mind and an overloaded nervous system is likely to produce stress hormones. In turn, our rate of breathing is likely to increase, with the result that we release excessive CO_2.

As Tess Graham writes:

It seems that today we can be so disconnected from our bodies and so taken in by technology that we do not think to look closely at ourselves for either the source of a problem or its solution.[26]

I must acknowledge that it was primarily the effect of an ageing body and not any accumulation of wisdom on my part that was the catalyst for me to slow down. However, I also realised that in order to effectively practise Buteyko breathing I needed to make it my goal to slow down generally.

By recording my pulse and my CP, I was able to gauge whether I was achieving my goal.

I have discovered that not only the quantity of food we eat but also the type of foods we eat may influence our rate of breathing. In fact, I believe that my diet was, on some occasions, responsible for my CP.

On 4 July 2012 I recorded an incredible CP of 75.5 seconds, the longest I have been able to hold my breath without feeling breathless.

I recorded the following:

CP of 75.5 seconds. Felt very calm and in need of little breath. Was it the effect of the Yulu tea? Dinner consisted of pork and fennel sausages, mashed potato, broccoli and carrots. No alcohol tonight. My next CP was 45 seconds.

I have found it very difficult to undertake reduced breathing if I have consumed alcohol. A pulse quickened by the effects of alcohol makes it almost impossible to slow one's breathing. It is equally difficult to reduce my breathing when my digestive processes are still at work. For this reason it is essential that at least a 2 hour interval should occur before embarking on reduced breathing.

I have found some foods more easily digestible than others. For example, red meat and seafood are not as easily digestible as pork and chicken. Processed foods can increase the rate of your breathing as all sorts of chemicals are introduced to your digestive tract. Fresh fruit, nuts, vegetables and meat should always be preferred, and foods that are high in sugar and high in starch should be avoided or at least limited as they cause quick changes to the blood sugar levels that lead to overbreathing.[27]

4

Liberation at last! – Freedom from the machine

At the conclusion of my six week Buteyko Breathing course on 22 May 2010 I was asked to fill out a questionnaire. In describing my condition and comparing it to when I started I wrote:

> *I have rediscovered the benefits of quality sleep. I have clarity of mind, my body functions like a well-oiled machine instead of a broken down toy. I am stronger, have lost weight* [28] *and am less tired – all whilst having an average of 6.5 hours sleep! I am in tune with my bodily functions and no longer take it for granted.*

The questionnaire also recorded the fact that at the start of my course my CP was 12 seconds. At the completion of my course my CP had increased to 50 seconds.

Following my first sleep study, you will recall that my sleep apnoea had been medically diagnosed as "severe". By the end of my course I believed that it was no longer severe and that sleep apnoea had in fact been largely nullified.

My belief was based on the greatly improved quality of my sleep and the results of my second sleep study.

My sleep was diagnosed as reasonable with a sleep efficiency of 84% and infrequent hypopnoeas.[29]

A comparison of some of my vital statistics is detailed in the attached Sleep Study Report, some of which is reproduced below:

Values	**2nd Sleep Study (20.05.2010)**	**1st Sleep Study (14.10.2008)**	**Average Values**
Weight	87 kg	95 kg	(78 kg)
BMI	27.8 kg/m²	31 kg/m²	(18.5–24.99 kg/m²)
Sleep Period	343 min	432 min	(Varies)
REM Sleep	**18.4%**	**21.3%**	**(20%)**
Sleep Efficiency	84%	81.8%	(85%)
AHI*REM	3.8 per hour	65.5 per hour	(< 5.0 per hour)
SaO_2% Lowest	93%	83%	(95%)
Average SaO_2 Desaturation	2%	4%	(Not applicable)

*The Apnoea Hypopnoea Index measures the combined number of apnoea and hypopnoea per hour. Significantly, less than 6% of the total number of apnoeas/hypopnoeas recorded during my first sleep study was recorded during my second sleep study.

Hypopnoea is defined as a greater than 30% reduction in airflow which lasted at least 10 seconds and was associated with a decrease of at least 4% in arterial oxyhemoglobin saturation.[30]

My results on the AHI were very significant *as less than 5 hypopnoea per hour is considered normal.* The following statistics are used in categorising the varying degrees of obstructive sleep apnoea:

- 5–15 hypopnoea is mild sleep apnoea

- 15–30 hypopnoea is moderate sleep apnoea
- Greater than 30 hypopnoea is severe sleep apnoea[31]

So my diagnosis based on the AHI criteria went from severe sleep apnoea to normal, that is, no sleep apnoea.

Interestingly, although not a single *obstructive apnoea* was recorded, there were three incidences of *central sleep apnoea* observed.

The difference between the two forms of sleep apnoea appears to be best explained by identifying whether the lungs make an effort to inhale during the period of absent airflow, or apnoea, (obstructive) or whether no effort is made by the lungs to inhale during the apnoea (central).[32]

Also significant was that:

1. My lowest oxygen saturation level (SaO_2) was at 93% – some 10% greater than recorded during my first sleep study.

2. My desaturation[33] level of my oxygen had improved by 100% when compared with my first sleep study.

I can now appreciate that if my oxygen levels in my haemoglobin were lower than what they should have been, then my tissues and organs were not receiving the required oxygen.

Shortly after my second sleep study I discussed the results with my sleep disorders specialist. It was a slightly odd discussion as he told me that there was a 60 second period where my airways were restricted even though I continued to breathe. Then my brain "kicked in" to open up the airways. I don't believe that my brain had anything to do with my airways being opened up, as it is my understanding that the increased CO_2 levels would cause the airways to expand.

But I had no doubt whatsoever that my health had improved vastly in the six week period after I had commenced the Buteyko breathing. And I felt great!

On 2 June 2010 my sleep disorders specialist wrote to my GP in the following terms:[34]

I reviewed Paul regarding his obstructive sleep apnoea. Paul continues to feel that symptoms are well controlled using the Buteyko technique. With this, he is feeling well through the day without significant sleepiness.

Paul's most recent sleep study performed with the mouth taped and using the Buteyko technique showed that his sleep apnoea is improved compared to previously. There are still episodes of hypopnoeas occurring during REM, but these are not associated with oxygen desaturation. Overall Paul's sleep apnoea is now of moderate severity, whereas previously this was severe.

As Paul's symptoms are well controlled using the Buteyko technique, and he [sic] sleep apnoea is not sufficiently severe for me to be concerned about ongoing cardiovascular effects, I am comfortable with him continuing to use this technique as treatment for sleep apnoea. Whilst I have not scheduled further appointments for Paul, he knows to return to see me if he has a recurrence of symptoms.

Fortunately I have not needed to return to my sleep specialist. Although I disagree with the diagnosis of my sleep apnoea as "moderate", I do not wish to become fixated on labels. What has occurred since that sleep study and is continuing to occur is that I am consistently enjoying quality sleep which enables me to pursue my daily activities (both work and social) without any of the deleterious effects associated with sleep apnoea.

The letter written to my GP by a sleep disorders specialist was significant in that it appeared to approve the Buteyko breathing, or at least openly acknowledge the benefits of Buteyko breathing being used to treat sleep apnoea.

Not that I needed such an acknowledgment – my improved condition was sufficient acknowledgment for me. My greatest pleasure lay in the independence I had achieved. That independence was exemplified by a freedom to breathe in a quiet and even manner without having to rely on an external regulator.

I readily accepted the need to rely on an external regulator to breathe whilst scuba diving. After all, I was entering a foreign environment where an aqualung was essential. But the same rationale did not sit well with me when I relied on CPAP to regulate my nocturnal breathing whilst I was in a very familiar environment – my own bed!

It has been reported that up to 90% of Australia's population has been at one time or another affected by sleep disorders and that up to 30% of the country's population is affected at the one time by sleep disorders. Those sleep disorders have been caused by a number of factors including sleep apnoea, narcolepsy, obesity and snoring (which often belies something more sinister).[35] It has further been reported that the annual cost of sleep disorders for treatment and lost productivity is $10 billion.[36]

In 2017 in Australia it has been estimated by a study conducted by Deloitte Access Economics that inadequate sleep has caused the following losses:

- $1.8 billion health system costs
- $17.9 billion productivity costs
- $40.1 billion wellbeing costs*

*Wellbeing costs estimated using World Health Organisation and Australian Government metrics which assess the non-financial costs of healthy life lost through disability and premature death from inadequate sleep and associated conditions.[37]

How effective is breathing retraining?

In 2012 the Buteyko Institute of Breathing & Health published a report (BIBH report)[38] based on:

- The findings of a survey conducted in 2010
- The collective experiences of Buteyko practitioners (67% of whom had been in practice for more than 10 years)
- More than 11,000 clients with sleep apnoea who had been treated
- Practitioners' responses from Australia, USA, UK, New Zealand and Canada

The purpose of the report was to ascertain how effective Buteyko breathing retraining was for clients suffering from sleep apnoea.

The report records that the Buteyko method was developed in Russia in the 1950s and was endorsed as a mainstream therapy for asthma in the Soviet Union in 1983. Following its introduction to Australia in 1990 the method was taught mainly to people with asthma. However, in the past decade, increasing numbers of people with sleep apnoea in Australia and overseas have attended courses in breathing retraining using the Buteyko method.[39]

The BIBH report[40] found that:

- Over 95% of participants showed significant improvement in sleep
- About 80% of participants were able to cease using their CPAP machine
- Symptoms such as snoring, headaches, restless legs, low concentration levels and decreased energy levels also improved in the majority of clients

The findings of the BIBH report have led to a call for clinical trials of the Buteyko breathing method in combating sleep apnoea to be held as soon as possible. There have been many clinical trials around the world conducted for asthma sufferers who had been taught Buteyko breathing.

The first such clinical trial in Australia was conducted in Brisbane at the Mater Hospital from January 1995 to April 1995. The results were published in the *Medical Journal of Australia* and showed that after twelve weeks participants had an average of 96% reduction in their bronchodilator (reliever) medication, an average 49% reduction in steroid (preventer) medication and an average 71% reduction in asthma symptoms.[41]

Breathing during exercise

As part of my Buteyko course I learnt to incorporate my reduced breathing into my exercise. I would inhale two-thirds of my usual breath down to my diaphragm and slowly exhale whilst I walked.

Keeping my walking evenly paced and my breath controlled and steady I soon experienced an astounding bodily response.

My breathing became more efficient. I felt my system "kick into gear" as if it had realised that this was all the amount of air it was going to receive and it had better process it efficiently. My heart was hammering away as my respiratory system functioned strongly.

At a later stage I used the method when running for short distances. As I was not a regular runner I did not persist for very long. But it was long enough for me to appreciate the method's potential.

Whilst training at the gym with my daughter, I decided to experiment with reduced breathing when on the treadmill. You may be familiar with the features on these machines. They have the ability to preset a program or you can manually input your own speed, degree of difficulty and duration. Each program calculates the number of calories you have expended and monitors your heart rate.

I experimented with the heart rate. After several minutes exercising on a simulated flat terrain whilst steadily breathing I took a mental note of my heart rate before switching to the method.

After a minute or so of reduced breathing I saw my heart rate fall significantly. The rate fell by between 4 to 6 beats per minute! At first I was sceptical and wondered whether there was something wrong with the machine. I repeated the experiment time and time again and my heart rate consistently fell when applying the method.

I have used the method when swimming (breaststroke) and playing tennis. Based on this experience I am able to vouch for the potential for athletes using the method.

In fact, I understand that my Buteyko teacher trained elite athletes who achieved outstanding results insofar as their personal best times were concerned.

The manner and frequency of my breathing had now begun to inform how I lived.

5

To be in control or to be controlled – that is the question

It is true that what matters most to us in life is worth fighting for. Whether it is values or qualities which we hold precious, nothing of any real and enduring value comes easily.

It takes a great deal of discipline to initially adhere to the reduced breathing exercises taught by Buteyko. As Professor Buteyko himself said:

> *My dear people; my method helps those who have will power. Those for whom the will is a weak spot will die of their illness.*[42]

Although these are very confronting words, they serve to illustrate for me two things. First, the life threatening nature of sleep apnoea and how it detrimentally affected the quality of my life. Second, the self-discipline required by me to consistently undertake the reduced breathing sessions.

Life was such a struggle for me that when given the opportunity to control my sleep apnoea I regarded the importance of adhering to the reduced breathing exercises as a matter of life or death. That may sound overdramatic. But the desire to live independently of CPAP and my determination to avoid ever again experiencing the debilitating effects of sleep apnoea were so strong that I applied the same level of committed discipline to the breathing exercises as if I had been told that certain death awaited me if they were not done.

Although I was often tempted to ignore the exercises, lie down on my bed, apply the mouth tape and try to sleep, I knew that only persistent diligence would result in my brain's respiratory centre being rewired.

We live in an age of entitlement and immediate gratification. We have been programmed by the technological age to expect that our cyber demands will be met and our tasks will be completed. If we are let down by technology we feel frustrated and to some extent helpless as we have been almost conditioned to expect that we cannot live without our smartphones, iPads, tablets and personal computers.

The instant access to knowledge and communication brings with it an expectation of immediacy. Unfortunately this cyber culture breeds or feeds impatience and is dismissive of reflection. It rewards speed, impetuosity and reflexive responses.

What is needed to successfully tackle breathing disorders is a toolkit which contains tools that are the antithesis of that found in the cyber environment. You will need to reflect on how you are breathing and how your body and mind are responding to your breathing.

You will need to exercise discipline in carrying out reduced breathing and have patience in dealing with the difficulties and setbacks you will experience as you make the transition from a rapid mouth breather to a controlled and regular nasal breather.

We also live in an age of the "quick fix". Have a headache? Pop a pill. Feeling anxious? Take a Valium™ or other sedative. Feeling depressed? Begin a course of antidepressants.

Sometimes these solutions are justified and may assist in the short term and may even be necessary in the long term. But there also may be a real risk that we are treating the symptoms rather than the cause.

I recall, when in my late thirties, I was accompanying a group of work colleagues one night who were entertaining a colleague visiting from Sydney. Much drinking was planned but I had a chronic and persistent headache. This occurred before my sleep apnoea had been diagnosed. When I mentioned to my colleagues that I doubted whether I could participate in the drinks planned for that night, the interstate visitor suggested that I take a particular analgesic containing ibuprofen[43] which would enable me to drink with impunity. He was right! The headache quickly disappeared and I was able to partake in the festivities without discomfort.

But I could hardly recommend an analgesic fix prior to consuming alcohol as a long term solution!

I hope that this discussion illustrates that it is a far more preferable and enduring solution to not only tackle the cause rather than the symptom, but also to do so in a manner in which you are in control.

Dr Buteyko, as a young doctor, suffered from hypertension. The increase in blood pressure manifested in a splitting headache and blood throbbed like a hammer in his temples. His pounding heart would be in pain and his right kidney would ache. His research had led him to suspect that deep breathing was the reason for his hypertension. So he began to breathe less. He told himself there were to be no deep inhalations or strong exhalations. He was to breathe only a little.[44]

What follows is reproduced below from Sergey Altukhov's book *Dr Buteyko's Discovery*:

> *He felt as if he was running out of oxygen. He wanted to open his mouth and swallow great gulps of air, but he restrained himself. A minute passed, then two, then three, and the miracle occurred. A true miracle. Buteyko's headache began to disappear and the pounding in his temples ceased. The pain in his heart subsided, leaving him feeling wonderfully relaxed. His aching right kidney felt as if it had been soothed with a hot compress.*
>
> *"It actually worked." Buteyko couldn't quite believe it. He deliberately took several deep breaths and his symptoms instantly began to return. He reduced the depth of his breathing, and the symptoms disappeared again.*[45]

I recall experiencing exactly the same blood pounding of the temples and searing headache pain (radiating from above my right eyebrow in a diagonal line over the top of my head). Nausea and stress accompanied my pain. I had been absorbed in studying some legal documents and my concentration must have caused my breathing to quicken over an indeterminate period of time.

I sat upright with a straightened back and deliberately and determinedly slowed my breathing rate. Both my inhalations and exhalations were slowed as much as I could without too great a discomfort. Within 5 minutes or so the pain had disappeared and I felt calm.

CO_2 affects the tone of smooth muscle in arteries and headaches can occur in response to changes in the tone – constriction and dilation – of arteries feeding the brain.[46]

Why did this happen to me?

It is only during the course of writing and researching for this book that I have been able to answer this question. Yes, I had always mouth breathed and yes, I had always had a deviated septum. So why didn't sleep apnoea rear its ugly head sooner?

I believe that our minds and bodies have a remarkable ability to adapt and cope with stress. But sometimes an incident or event will be the "straw that breaks the camel's back". For me, it was the death of my mother and, within a further 16 months, the death of my father.

All other aspects of my life were "apparently" being kept in check. But the grief I experienced from these events increased my dysfunctional breathing. As a consequence my CO_2 levels decreased as did the quality of my life.

The "breathing" settings within my brain needed to be reset. Old habits needed to be discarded and new habits needed to be embraced so as to allow an increase in my CO_2 levels.

Ironically my mother suffered a stroke when she was only 43 years of age. Her undiagnosed sleep apnoea may well have played a part in her stroke due to high blood pressure caused by her sleep apnoea. Also the lack of oxygen during an apnoea results in the body adopting a "fight or flight" response. Consequently this leads to blood clotting and blood clots in the brain can cause a stroke.[47]

As a stroke victim it is likely that my mother was given a premature death sentence. It may assist you in appreciating what follows if you refer back to my earlier discussion in Chapter 4 of the comparative values of my two sleep studies.

The total number of apnoea/hypopnoea that my mother would have endured per hour (in other words, her Apnoea Hypopnoea Index [AHI]) would likely have placed her in at least the *moderate* if not *severe* category of sleep apnoea sufferers. In fact, Swedish researchers (who followed 132 stroke patients over 10 years) found that those with an AHI of 15 or greater (that is in the moderate or greater category) were 76% more likely to die earlier.[48]

As sleep apnoea is a treatable disorder, it is realistic to expect that the life of a stroke victim who suffers with sleep apnoea may be extended by addressing his or her sleep disorder.

How can you ensure this doesn't happen to you?

This is a challenging question and of course I can only discuss the lessons learnt from my own experience. But the following tips appear to have universal application:

Maintenance

Just as we carry out maintenance for our modes of transport (bicycle, motorcycle, car or boat), so we should maintain our bodies. Our body is our mode of transport through life. It is our human-powered vehicle.

Just as our body needs food, air and water, it is vital to be aware of how our body receives the food, air and water. We should not eat or drink excessive amounts of food or water. If we eat excessively we risk serious long term health complications associated with obesity. If we drink excessively we risk depleting our body of important minerals and electrolytes which assist in transmitting electrical impulses for the proper functioning of our heart, nerves and muscles.[49]

Overhydration was experienced by my brother whilst trekking in Papua New Guinea on the Kokoda track several years ago. He was very fit and was in the lead group. But he succumbed to the temptation to drink far too much in anticipation of thirst. He also contracted a bug and sadly had to be airlifted by helicopter to Port Moresby after completing two-thirds of the trek.

And so it is with the amount of air we breathe. By only breathing through our nose and practising reduced breathing (under guidance from an experienced breathing retrainer) we will go a long way to ensuring that we breathe the right amount and in the right manner.

In controlling my sleep apnoea I have reached a stage where I periodically check my CO_2 levels by performing a CP. I may undergo a reduced breathing session if I have a headache or a headache reminder. Also if I am feeling "low" or "flat" a breathing session will always assist by centring my focus by relaxing and becalming me.

Conscious Exercise

Exercising in a conscious manner means being aware of your pattern of breathing and the volume of air inhaled.

It is not essential to practise the method discussed towards the end of Chapter 4, as long as nasal breathing is practised in an even and controlled manner.

Sleeping Right

Never sleep on your back. Always sleep on your side (preferably your left side) with your head in a slightly raised position. You can use two pillows as I do. If you have difficulty in keeping your mouth closed, you can tape your mouth. I use Leukopor® tape. After trying a few brands I prefer this product as it is a hypoallergenic tape, is comfortable, has good adhesion and actually "breathes". The tape is permeable to air and moisture, comfortable and can be easily torn from the roll to whatever length you wish.

Eating Right

The type and quantity of food we eat can make our respiratory system work harder. As I have learnt there are already too many other stress factors in ordinary life which place pressure on our breathing.

It is also important to eat slowly. Don't rush, as you want to avoid allowing air to be mixed in with mouthfuls of food. Also you should eat at least a few hours before sleeping.

Finally caffeine and alcohol should also be avoided within a few hours of sleeping.

Knowledge is power and the power comes from being able to exert some control over our body so that it performs as it should.

To conclude

I genuinely hope that after having read about my journey and after having reflected on my experiences and thoughts that this book will herald a new beginning for you. Hopefully it will motivate you to critically examine the quality of your life and consider the possibility of change.

I believe that once you have gained an understanding and true appreciation of the vital need to produce and maintain an adequate blood level of carbon dioxide, you will believe.

But I am also convinced that you will only attain this belief once you have actually experienced and enjoyed the amazing benefits of an improved quality of life. This is the "rub" or the "Catch 22", as I believe that only through deeds will thoughts follow. It is like romantic love – you cannot truly know it until you have experienced it.

Or as a wise and influential thinker and philosopher once said,

"I hear and forget. I see and remember. I do and I understand."[50]

Awareness, or mindfulness, of our breathing is essential. Recognising how everyday emotional and stressful states affect the volume and rate of breathing is essential to achieving a state of relaxation. By focussing on lengthening the inhalation and exhalation of the breath and breathing deeply into your abdomen, your respiratory rate will decrease and you will begin to relax.[51]

It is only after you have felt the strength of newfound energy and listened to the quiet of your body's efficiency that you will appreciate carbon dioxide as an elixir of life.

Only then will you be ready to embark on a quest to keep the genie in the bottle and not let this most precious of gases escape from within your temple.

As I have been fortunate enough to have experienced.

6

Speaking in Tongues

For the past eight years or so I have almost always slept with my mouth taped at night.

Sometimes I would forget to do so, especially after a few too many drinks! Sometimes I would simply hope my mouth would remain closed during the night. After all, I managed to enjoy an afternoon nap without taping my mouth. On these occasions I awoke an hour or so later with a moist and closed mouth.

But sleeping at night is very different to napping during the day. During the night, we experience various stages of sleep including REM. In my experience, drawn from two sleep studies, I found it more likely that apneas and hypopneas were likely to occur during the REM stage of sleep.

Even though mouth taping greatly assists me in achieving good quality sleep I have sometimes yearned for a completely natural state where I no longer apply mouth tape during the night. However, on those occasions when I have retired to bed after a little too much alcohol I have discovered to my delight that mouth taping erases any headache I was suffering immediately prior to sleeping. I often marvelled at going to sleep *with* a headache and waking *without* one!

I wondered what other alternatives existed to prevent my mouth from opening during sleep. My ambivalence towards mouth taping continued for some time until a series of fortunate events occurred which potentially revealed a missing piece from the jigsaw of my respiratory health.

For the past few years I often thought of training as a Buteyko practitioner. I enjoyed an opportunity of discussing the health benefits of the Buteyko Breathing Method (BBM). Sometimes my enthusiasm was challenged by a listener's arched eyebrow or sneer. "How can my health improve just by changing how I breathe?" "Why should I let someone teach me how to do something I have been doing all my life?" These questions reminded me of my initial cynical response when I was first introduced to the BBM.

I was reminded once again that a person who breathes in a dysfunctional manner must first experience its benefits before (s)he can understand and appreciate the BBM.

The Revelation

I attended a BBM training course in Sydney from 22 - 25 October, 2016 for aspiring Buteyko Practitioners. The course was conducted by Patrick McKeown, a world renown Breathing Educator who had trained with and was accredited by Professor Buteyko. During the course one of the trainees, Bridget Ingle, detected my tongue tie. She was an astute orofacial myofunctional therapist and a lactation consultant. She observed it when I spoke and swallowed.

Myofunctional therapy, also called orofacial myology, is the neuromuscular re-education or re-patterning of the oral and facial muscles. It uses specific therapeutic techniques over a period of time to assist in proper functions, growth and development, and proper aesthetics.[52]

I knew that I had a tongue tie (or *ankyloglossia* as it is medically known) and recall my parents telling me when I was a young boy that the family Doctor did not recommend any corrective action as my speech was unlikely to be adversely affected by it. I believe this diagnosis was more reflective of the state of medical knowledge at the time rather than due to any misdiagnosis.

Why was a tongue tie relevant to my breathing?

In a ground-breaking study on 27 March 2016 led by Professor Christian Guilleminault, a world leading sleep medicine specialist, he investigated whether there was an association between a short lingual frenulum and obstructive sleep apnoea syndrome (OSAS) in children.

The study involved 63 children with a short frenulum and 87 children with a normal frenulum aged between 3 to 12 years who were referred to their sleep clinic for sleep disorders. Of those children with a short frenulum 96% exhibited fatigue and 80% had a high and narrow upper palate.

It was stated that *a short lingual frenulum has been shown to lead to mouth breathing and to abnormal development of the oral cavity, increasing the risk of upper airway collapsibility during sleep.*

Importantly it was noted that *genetic mutations present at birth may not impact the upper airway until later in life. Environmental factors may be present for a long time before the onset of clinically observed sleep disordered breathing, which may result in clinical symptoms only recognised during adulthood despite the presence of anatomical and/or functional anomalies during childhood.*

The results of prior studies observed both mouth breathing and abnormal anatomical findings at the time of recording and diagnosing OSAS. The study concluded:

When considering results of several of our investigations performed in children a pattern emerges: a dysfunction early in life involving abnormal nasal breathing, sucking and masticating leads to progressive dysmorphoses favouring increased collapsibility of the upper airway during sleep, which worsens with ageing and leads to the development of sleep disordered breathing over time up to adulthood. [53]

One afternoon, at the end of the day's training, Bridget and her colleague asked me if I would perform a series of tongue movements as part of a "kerbside consultation". I was curious and interested to learn more as my tongue had not previously featured in my breathing re-education.

When asked to perform a tongue thrust my tongue tottered timidly forward only to be reeled back into my mouth like a coveted prize marlin! So strong was the pull of the tongue tie that it overcame any mental command. When I attempted a lateral thrust my tongue struggled like an exuberant Labrador pulling against its leash. The Newtonian law of physics did not seem to apply to my retarded tongue as every forward action did not have an opposite and equal reaction. Perversely it had an opposite and unequal double reaction! The most disappointing aspect of these exercises was my tongue's inability to reach my upper palate.

My physicians were wide eyed and exchanged knowing glances. Bridget's colleague asked me if I had recently suffered a stroke! I shook my head and smiled as my exhausted tongue lay panting on the floor of my mouth.

Bridget showed me what exercises my tongue should be able to perform. Her tongue was of an Olympic strength and flexibility. I quickly realized what an underperforming tongue lay in my mouth.

The BBM course taught me that people who breathe through their nose normally have three quarters of their tongue postured up into the maxilla (top jaw) pressing against the upper palate. The tip rests gently behind the top front teeth. When the tongue sits right up behind the front teeth, it is maintaining the shape of the top jaw every time you swallow.[54]

While your tongue is in this position it is impossible to breathe through your mouth. It all made sense to me now. This is one of the main reasons why I became an unconscious mouth breather. As my tongue was unable to rest against the roof of my mouth it lay on the floor of my mouth facilitating a large volume of unfiltered air down my throat and into my lungs.

Although I had converted to nasal breathing eight years ago my tongue tie prevented my tongue from adopting the above posture. My tongue either sat on the floor of my mouth or floated rather insecurely within the mid space of my mouth. I was told by Bridget that I needed to prepare my mouth and tongue for the tongue tie release or, what is medically known as, the frenectomy. She also recommended a Dentist who specialized in this area and some treatment from an Osteopath to support all the facial and bodily changes that will occur. Bridget described the neck, jaw, tongue triangle as connected by a series of levers and pullies. When something is short, something else is long.

Pre-operative

Bridget prescribed two weeks of tongue strengthening exercises to prepare for the frenectomy. These exercises would enable my tongue to begin to adjust with being "let off the leash". I understood that a significant release of tension and expanded range of movement would inevitably occur.

I made an appointment on 8 December 2016 with the Dentist.

I saw Lauren, a registered Osteopath, on two occasions prior to the frenectomy. Lauren observed both tension and or tenderness in my:

1. shoulder blades, upper back and neck
2. muscles over the upper segments of the neck and in under the jaw
3. *scalene* muscles – defined as any of four pairs of muscles extending from the cervical vertebrae to the second rib; involved in moving the neck and breathing [55]
4. cervical and pectoral fascia
5. throat
6. *hyoid* bone – defined as a U-shaped bone at the base of the tongue that supports the tongue muscles.
7. *masseter* – defined as a large muscle that raises the lower jaw and is used in chewing[56]

8. *pterygoid* muscles (between the teeth at rear of jaw) and *genioglossus* muscle fibres (under the tongue)

My lingual frenulum or tongue tie was also very tender to touch.

Interestingly Lauren also noted that breathing through my nose appeared to require some physical effort whilst talking. Specifically, she saw tension in my scalene muscles and heaving of my chest when I inhaled through my nose prior to speaking.

Lauren also observed stiffness in my chest motion on respiration and on passive motion testing through the costovertebral joints (that is, where a rib connects with a vertebra) and the thoracic spine itself.

It really was a revelation to learn what the trained eye saw!

Finally, my upper neck range of movement was poor and my tongue motion was noted as being restricted with reduced elevation.

On both occasions, I was prescribed exercises which included neck stretches, jaw and spinal mobility exercises in addition to the tongue strengthening exercises I was performing. I was encouraged to continue practising certain yoga poses such as the cobra, downward dog, cat and child's pose as they assisted with the prescribed exercises.

The Release

It wasn't until I was in the Dentist's waiting room that I became anxious. My heart rate and number of breaths increased. My chest heaved. I did not relish the prospect of my tongue being cut or lasered whilst I was awake. I recall thinking that I didn't really have to go through with this if I didn't want to. I imagined standing up, crossing the floor and walking out the door. I shut my eyes and began practising reduced breathing. Within minutes I was able to regain a sense of calm and focus.

By the time I was sitting in the Dentist's chair I was fascinated to learn of the procedure. It involved a local anaesthetic needle to either side of my tongue. Once the anaesthetic had taken effect a laser was applied to sever the frenulum. It sounded simple enough.

However, I was not prepared for the smell of my burning tongue. It was a surreal experience. But I soon saw the funny side of the operation and began making mind jokes about my barbecued tongue! The entire procedure was over in a matter of minutes. A mirror was placed before my open mouth. I raised my swollen but liberated tongue and saw a gaping wound. My Dentist was pleased with the result. He endorsed the continuation of my exercises following a recovery period of 24 hours. He prescribed an additional exercise. It bizarrely involved repeatedly stroking my open wound with my index finger four times each day for ten repetitions.

The purpose of the finger stroking was to prevent the wound from healing, reattaching or scarring while I performed my rehabilitation exercises. Deb drove me home. I started to feel tired as the anaesthetic wore off.

I clumsily managed to drink several spoons of some lukewarm soup. As fatigue set in I lay down on my left side conscious of my tongue resting, for the first time, on the roof of my mouth.

Post- operative

Over the next few days I was both dismayed and exhilarated at the newfound strength and flexibility of my tongue. Dismayed at how quickly my tongue tired while eating and speaking (of which I did little) but exhilarated with the range of movements which were now starting to make themselves known. Chewing more than normal was recommended. Although this made eating a more time consuming and tiring task I was enabling greater nutrients to be extracted as my saliva broke down the vitamins and minerals in the food. I understand that eating slowly also enabled hormones to relay a message to my brain that I had eaten and that nutrients were being absorbed. This process prevents you from overeating.[57]

There is a respiratory benefit from eating slowly. If you eat quickly and gulp down your food you are more likely to take in gulps of air through your mouth at the same time. This unsavoury habit will not only bloat your stomach but also cultivate dysfunctional breathing. By contrast, regular nasal breathing whilst you slowly chew your food will assist the pre-digestive process that takes place in your mouth enabling easier swallowing and improved digestion.

I saw Lauren on 12 December 2016. Although the frenectomy was only performed 4 days ago she noticed a reduction in soft tissue tension in:

- my cervical region
- the erector spinae muscles
- scalene muscles
- anterior fascias around the throat and hyoid

A greater range of tongue elevation and motion was observed. I was encouraged to continue using familiar yoga poses to open the anterior fascial planes.

On 30 December 2016 I slept without taping my mouth. Apart from the odd occasion when I had forgotten to apply the tape before falling asleep – only to awake in a groggy and unrefreshed state – this was the first time I can recall consciously doing so in almost 8 years.

Unusually I awoke at about 2.00am thinking it was time to arise. After a short time I fell asleep with my tongue planted firmly under the roof of my mouth and positioned behind my upper front teeth.

Upon awakening the next morning my mind was clear and alert. Deb had seen me asleep with my mouth firmly closed and quietly breathing.

On 6 January 2017 Lauren observed that my wound had healed well and the resting position of my tongue was improving in its contact with the palate. I had noticed that it was relatively effortless to breathe through my left nostril. My deviated left nasal septum was less taut. I noted an easier air intake for each inhalation.

Lauren observed upon examination:

- very soft and relaxed anterior neck and throat tissues in both upright and reclined postures
- improved upright posture with less forward head carriage
- jaw free of tension, very mobile hyoid, mouth closed with comfortable nasal breathing

- mandible was without deviation
- softer pterygoids and not tender at all upon touch
- Relaxed genioglossus muscles and not tender at all upon touch although the intrinsic tongue muscles were tender upon touch

Over three nights from 9 January 2017 I slept well without a taped mouth. I awoke each morning with a moist mouth and feeling rested and alert. My energy levels were good throughout each day. Clearly, I was attaining an adequate level of blood oxygen saturation.

Lauren considered that the added pressure from my tongue resting against the upper palate and the greater blood flow together with a relaxation of the pterygoid muscles was likely to have contributed to the improvement of my deviated nasal septum. She further commented:

I believe your dedication and awareness of the breathing and yoga principles have helped you a lot. I have been quite astounded at the rate of change in the anterior neck and hyoid and through the fascias of the anterior body[58]

I continued to practise the post-operative tongue and neck postural exercises recommended by Bridget as part of my journey in retraining my brain and creating new muscle memory.

The flexibility and strength of my tongue continued to astound me. I now saw that my tongue was previously in the equivalent of a straight jacket. It now sits proud and strong in contact with my upper palate ready to uncoil itself for the various tasks of eating, drinking and talking.

But it plays no role in my breathing. Nor, as I have discovered, should it play any role in your breathing.

From its ergonomically alert position it facilitates quiet and efficient nasal breathing. As previously alluded it is impossible for me to mouth breathe while my tongue rests against my upper palate. Upon speaking with many people who naturally had their tongues resting against their upper palate I soon realized that almost all were unaware of their tongue position. These facts provided me with a powerful insight regarding the manner in which we should breathe. It was my Damascene conversion.

I gained enough confidence to sleep for 5 consecutive nights without taping. I felt refreshed and knew that I had slept well. But during the next night I found myself on my back with my mouth open snoring indignantly. When switching from side to side during sleep I sometimes found myself being caught "in the middle" when on my back. Fatigue caused me to linger longer while supine. What should have been a quick pitstop on my journey from one side of the lateral border to the other became a prolonged recovery stop. My sojourn ended when I was either woken by Deb or my snoring interrupted my supine sleep.

On these occasions the quality of my sleep suffers and I stay in bed longer than usual to "make up" for lost sleep.

Having a rough night, as just described, shakes my confidence to again sleep at night without taping. If I wake unrefreshed it does not matter how "normal" or "natural" it feels the next morning upon waking with a tapeless mouth. The effects of poor quality sleep and the consequential struggle the next day is a powerful incentive for me to tape my mouth to ensure a quality night's sleep.

Why do I sometime sleep well and other times sleep poorly without taping? I suspect the higher my CP the more likely I am to enjoy a good night's sleep without any assistance. But if my CP is lowered due to a lapse with my diet and exercise regime, thereby stimulating and increasing the rate and volume of my breathing, I am more likely to lean on the crutch that is provided by mouth taping.

It is an ongoing challenge but one which is infinitely more appealing than the challenges which faced me 8 years ago.

7

What does it all mean? Or "less" means "more"

In my quest to understand why I now have more energy during the day and why I am now able to enjoy quality sleep at night I need look no further than the role played by CO_2 in my breathing.

Just as the beating of our heart is an automatic process, so too is the rhythm of our breathing "which occurs at least 20,000 times each day".[59]

If our heart is irregular or defective in some way and is allowed to go unchecked, it is bound to have a deleterious effect on our lifestyle and general health. Similarly one can easily accept that if our breathing is irregular or dysfunctional it too, if allowed to continue unchecked, will disrupt our body's internal equilibrium.

The main purposes of breathing are threefold[60], namely:

1. To supply our body with sufficient oxygen.

2. To remove excess CO_2.

3. To maintain a constant pH in our bloodstream.

It is common knowledge that the atmosphere we breathe has 21% oxygen and contains only a negligible amount of CO_2 at 0.04%. But what may be uncommon knowledge is that the gas mix we need to maintain in our lungs is 14% oxygen and 6% CO_2 within the alveoli that grow from the bronchial tree in our lungs.[61]

But if we overbreathe or hyperventilate, we will rid ourselves of too much CO_2 with a consequentially detrimental effect on the pH level of our blood and on our ability to transfer oxygen to our body's tissues. The latter effect is known as the Bohr effect on which I will later expand.

It is commonly known that the pH of a substance is the assessment of how much acid or alkaline is within that substance.

Breathing and the pH of arterial blood are reliant on each other to a large extent because breathing plays an important role in regulating the pH of the blood and the pH of the blood reciprocates by regulating the breathing.

If the pH of the blood alters due to abnormal breathing patterns, a person is said to have either "Respiratory Acidosis" or "Respiratory Alkalosis".[62]

The key goal of reduced breathing, as taught by Buteyko breathing, is to increase the level of alveolar CO_2 in the lungs. A reasonably accurate way to measure the CO_2 level, as taught by Buteyko breathing, is by recording the passage of time from the holding of your breath until you first experience the need to breathe. This event, you will recall, is known as a CP (or control pause). A CP of 40 seconds is equal to an alveolar CO_2level of 5.5% and a pulse rate of 70 beats per minute. This is regarded as the International Optimum level. Note that the Buteyko Optimum level is different, with a CP of 60 seconds, which is equal to an alveolar CO_2 level of 6.5% and a pulse rate of 68 beats per minute.[63]

Once you have attained a reasonable amount of alveolar CO_2 in your lungs, you will be assisted with the regulation of your breathing so that your body can perform its internal functions efficiently.

It has been my experience that when I attain a CP of 40 seconds my breathing is quiet and regular. My nasal passages are dilated and I have a general feeling of inner well-being and calmness. Perhaps, at that point, my body had achieved homeostasis?[64]

During my Buteyko course it was explained to me that the reason I experienced a pause in my breathing during sleep was that my lungs were waiting for the CO_2 levels to increase before the urge to breathe was triggered.

Our breathing changes its volume automatically in such close adjustment to the amount of CO_2 produced in the body that the alveolar air is kept constant in this respect. It is the chief respiratory hormone of our body.[65]

It is important to restate the scientific fact that within our body's internal environment or within the alveolar spaces of our lungs the alveoli contain approximately 6% of CO_2.

Scientists claim that millions of years ago, before blue-green algae grew in the sea and forests covered the land, air on Earth was quite different to today. It was similar to the atmosphere in the alveoli, meaning that a primeval–type atmosphere is retained in the lungs.

The developing baby in the womb grows in an environment which has a higher level of CO_2 and a lesser level of oxygen than the atmosphere in which a newborn baby finds itself.[66]

The Bohr effect

Haemoglobin is defined as "a red oxygen-carrying substance containing iron and present in the red blood cells of vertebrates".[67]

The tripartite relationship between oxygen, CO_2 and haemoglobin is critical to our healthy breathing.

In the early 20th century, Danish scientist Christian Bohr discovered that CO_2 pressure affects the ability of haemoglobin to carry oxygen.[68]

Low pressure of carbon dioxide means that oxygen is retained by haemoglobin and high pressure means that more oxygen is released to tissue cells.[69]

The body is like a well-balanced machine – when it is working hard and the tissue cells need extra oxygen, it is simultaneously making an abundance of CO_2 and heat, thus ensuring a steady flow of oxygen to the tissues.[70]

Given that it is the CO_2 level which determines the release of oxygen from the haemoglobin in our blood, it would appear to follow that with a greater level of carbon dioxide in the lungs, the greater is the release of oxygen to our cells and tissues.

A problem can occur, however, when breathing is increased to the level where too much CO_2 is exhaled. This compels blood to become more alkaline than normal and haemoglobin becomes "stickier", retaining oxygen instead of releasing it.[71]

So it would conversely appear that a lesser level of CO_2 in the lungs means that a lesser amount of oxygen is being released to our cells and tissues.

If insufficient oxygen is being transferred to our cells and tissues, we tire easily and ache. I recall that when my sleep apnoea was at its worst my headaches were accompanied by aching limbs and a feeling of listlessness. My energy levels were low and the sensation was reminiscent of a bout of flu.

It was not unusual for me to be catapulted out of bed at night with excruciating cramps in my calves or thighs. The thigh cramps were the worst. It was almost physically impossible to stand up and place weight on my leg so as to free myself from the painful effects of the iron-clamped cramp.

When tissues do not receive sufficient oxygen, lactic acid is produced by and accumulates in the tissue.[72]

So it appears that the less we breathe the more our prospects are enhanced of our breathing reaching optimum efficiency.

When NO means YES

Since the original publication of this book there have been significant developments in scientific research into the functioning of our respiratory system including the role played by the gas Nitric Oxide ("NO").

Although NO is known as an atmospheric pollutant, like CO_2, it is produced in several areas of the body and serves a number of useful functions. It is produced inside the walls of the blood vessels and it reverses the build-up of cholesterol and plaque in the arteries, ensuring that blood vessels stay open. It helps our immune system and assists with brain cell signalling.

NO has distinct functions in the cardiovascular system, the nervous system, and the upper and lower airways.[73]

Scientific research into NO has been evolving particularly over the past three decades since 1992, when NO was named "Molecule of the Year" by the Journal *Science.* In 1998, Dr Louis Ignarro and two other US scientists were jointly awarded a Nobel prize for their discovery of NO as a key messenger in signalling problems in our cardiovascular system.

One of the benefits of NO which has received the most attention is its ability to relax the smooth muscle layer in blood vessels. The dilation of the blood vessels (or vasodilation) increases cerebral blood flow and oxygen to the brain, decreases blood pressure and relieves angina. In addition, it has other benefits including lowering serum cholesterol levels and acting as an anticoagulant.

NO is perhaps commonly known for its role in Viagra as an important mediator by enhancing blood flow in penile erection during sexual arousal.[74]

A recent study at Case Western Reserve University School of Medicine in Cleveland, Ohio, USA resulted in the discovery that NO is the third gas involved in our respiratory cycle together with oxygen and CO_2.

The study showed that haemoglobin also needs to carry NO to enable blood vessels to open and supply oxygen to the tissues. The researchers found that NO controls the release of oxygen from red blood cells into the tissues that need it. There was an experiment with mice that were unable to carry NO in their red blood vessels. It was found that although their red blood cells were able to carry a full load of oxygen the mice could not unload it to their tissues.[75]

A Conundrum is created

Is it CO_2 or is it NO which facilitates the release of oxygen from haemoglobin to the cells and tissues of the body? Is it both?

You will recall from our earlier discussion in this Chapter, where a high pressure of CO_2 exists, the Bohr effect enables haemoglobin to release oxygen from the cells to the tissues in the body.

Of course, I have no answer to this apparent conundrum. However, it may be that for optimum delivery of oxygen to the cells, our body needs both CO_2 and NO.

How one breathes plays a critical role in the production and presence of NO within the upper and lower airways and CO_2 throughout the body.

But the answer to the conundrum may not be immediately necessary as there are so many benefits, other than those already mentioned, that can be attributed to NO....

Our respiratory system benefits by the production of NO in our paranasal sinuses and the nasal cavity. Those benefits include a defence in the upper airways against the growth of bacteria, fungi and viruses. It is suggested that NO may help to keep the sinuses sterile under normal conditions. It has also been shown that nasal breathing reduces pulmonary vascular resistance and improves arterial oxygenation compared with oral breathing in subjects without lung disease.[76]

Importantly, for nasal breathers, the nasal NO concentrations are higher at lower flow rates.[77] Therefore, it is important to breathe lightly and slowly in order to harness an optimum quotient of NO.

In the context of sleep apnoea, as NO is inhaled with each breath through the nose into the throat *en route* to the lungs, it plays a role in assisting the upper airway muscles keep the airways open. If insufficient quantities of NO are carried with each breath, the muscles in the throat may not work hard enough to maintain an open airway. If the airway collapses, the individual stops breathing. The gas transmits messages between the nose, throat muscles and lungs.[78]

In fact, it has been asserted that both NO and CO_2 would appear to have roles in communication between the upper and lower airways. The current evidence indicates that optimal management of disease processes in both the upper and lower respiratory airways needs to consider a "unified airway" model as nasal breathing equates to more efficient O_2 extraction and CO_2 excretion during exercise.[79]

It has been suggested that a reduction in inhaled nasal NO might contribute to the negative effects caused by oral breathing that occur during sleep disorders.[80]

In Chapters 2 and 5 I briefly mentioned the "fight or flight" response. Physiologically this occurs when the sympathetic nervous system is activated. During this stage, heart and respiratory rates increase, digestive secretions reduce, and blood vessels constrict. Consequently, the rate and volume of air breathed will increase. We may also hold our breath.

Conversely the relaxation response occurs when the parasympathetic nervous system is activated. During this stage, the heart rate slows, digestive secretions are stimulated, and blood vessels dilate. A study in 2005 by Dr Jeff Dusek and others investigated whether the relaxation response was mediated by NO. The depth of the relaxation response was found to be associated with increased concentrations of fractionally exhaled NO. The study concluded that the relaxation response may be mediated by NO helping to explain its clinical effects in stress related disorders.[81]

In 2010 a study involving 15 healthy volunteers, comparative lung distribution of blood was tested in circumstances where the subjects, while sitting in an upright position, were inhaling:

1. NO free air through oral breathing

2. NO internally enriched air through nasal breathing

3. NO externally enriched air through the mouth

One of the authors (Lundberg) had previously demonstrated that auto inhalation of NO from the nasal airways improved arterial oxygenation and reduced lung vascular resistance[82] but the effects of NO on distribution of lung blood flow had not previously been investigated.

The study established that, compared with oral breathing, nasal breathing resulted in a net increase of 24% blood flow to the apex area of the lungs. Similar effects were obtained with NO externally enriched air breathed in through the mouth.

Distribution of blood is important not only in terms of gas exchange but also for our defence against infectious diseases as poorly perfused lung areas are more susceptible to pulmonary infections.[83]

The authors of this important study demonstrate that nasal breathing counteracts the gravitational effects on pulmonary blood flow in the upright position by redistributing blood to nondependent areas of the lungs including the apex area.

As a fascinating aside, the authors propose that the results of the study provide evidence that the substantial production of NO in the upper airways by humans may be an important part of our adaption to life on two legs to improve the pulmonary blood flow distribution and gas exchange, despite the influence of gravity and to provide some protection against infection.[84]

Conclusion

Nasal reduced breathing allows pooling of NO in the paranasal cavity thereby resulting in NO enhancing the benefits obtained from a healthy blood level of CO_2.

If we breathe less and slow down our nasal breathing NO may accumulate in a greater concentration and may combine with CO_2 to oxygenate our tissue cells.

In summary, the multiple benefits we gain from the body's production of NO strengthens my conclusion that we should only practise nasal breathing (as opposed to mouth breathing) both at rest and during light to moderate exercise.

Hyperventilation – the enemy within

Before discussing why hyperventilation is so destructive it is helpful to consider the following definition of hyperventilation:

A pulmonary ventilation rate that is greater than metabolically necessary for the exchange of respiratory gases. It is a result of an increased frequency of breathing, an increased tidal volume or a combination of both, and causes an excessive intake of oxygen and the blowing off of carbon dioxide.[85]

I have no doubt whatsoever that I was hyperventilating when my sleep apnoea was at its worst. I would often sigh involuntarily and yawn regularly. This occurred while I had a feeling of breathlessness. I didn't feel as though I was breathing in enough air. It was a very frustrating and dispiriting experience.

The BIBH report refers to the research of Professor Konstantin Buteyko into disordered breathing patterns. He found that chronic hyperventilation was the most frequently diagnosed breathing disorder. He also found that intermittent hyperventilatory breaths (for example, sighing, yawning and gasping) were also common, often appearing in addition to a pattern of chronic hyperventilation.[86]

Buteyko also found that hyperventilation was not always apparent – either to the patient or to the doctor – and he called this condition "hidden hyperventilation".[87] Audible hyperventilation is something that we don't always experience, particularly if a person is a quiet (although rapid) breather. Unless the heaving of one's chest is detected the hyper-breathing may go unobserved.

At night hyperventilation may also go undetected unless it manifests in snoring. It is my understanding that snoring is the body's attempt to slow down excessive breathing which may lead to hyperventilation.

Hyperventilation has been linked as a contributor or cause of sleep apnoea. An excessive exhaling of CO_2 leads to hypocapnia.[88] As discussed previously, the Bohr effect will result in less oxygen being released from haemoglobin which will, in turn, detrimentally affect the oxygenation of the tissues and cells.

The BIBH report claims that several studies suggest an association between hyperventilation, decreased CO_2 levels and apnoea. Research conducted in 1994 shows an occurrence of either apnoea or hypopnea following voluntary overbreathing in conscious humans.[89]

There is also a physical effect of inhaling an excessive volume of air. The walls of our throat will narrow and can actually collapse. The faster the flow of air through the airway, the greater the negative pressure – vacuum or suction force – created on the walls of the passage (the Bernoulli effect). In other words, the airway walls move closer to each other the faster you breathe.[90]

CPAP is designed to counter the Bernoulli effect by regulating the amount and frequency of the air inhaled so that your throat muscles will stay relaxed and your airways open.

So we can conclude from our physiological profile that we need a certain amount of CO_2 to process oxygen efficiently into our bloodstream and tissues.

If we overbreathe in an attempt to obtain more oxygen (and in doing so blow off CO_2) the reality is that we will actually obtain lesser benefits from oxygen.

My wife has suffered with an autoimmune disease[91] for over 30 years. It has often interfered with her sleep. She is a diminutive and elegantly boned woman who breathes quietly. But at night I have often heard her breathe rapidly. At times she would inhale up to three times the number of breaths I would inhale per minute.

When I asked my wife how she slept it was not unusual for her to reply that her sleep "was a bit restless". This phrase was code for her having had a bad night's sleep after too much breathing with too little oxygen exchange. She rarely complained of her condition despite its debilitating effects. Apart from being one of the most angelic women I have ever known, she is also one of the toughest women I have ever known.

It is interesting to reflect on how an excess of anything is not to our benefit. The Greek philosopher Plato once said:

Excess generally causes reaction, and produces a change in the opposite direction, whether it be in the seasons, or in individuals, or in governments.

I am sure that Plato would have little difficulty in applying his statement to a reduction in breathing in response to an excess of air thereby giving credence to the maxim "less is more".[92]

Appendix 1

Sleep Study 14.10.2008[16]

Vaucluse Hospital - Sleep Disorders Centre

Patient Name: Paul RODRIGUEZ
Date of Birth: 4/1/60
Weight: 95 kg
BMI: 31.0 kg/m²

Study Date: 14/10/2008
Sex: M
Height: 175 cm
UR: 048189

Diagnostic Sleep Study Report

Observations: PR = 96 ; BP = 115/77

Medical HistoryHC
The patient's BMI was 31.0 kg/m^2, which is in the obese range.

Sleep architecture
The total recording time was **474.5 min**, and the total sleep time was **388.0 min**. There were four periods of REM sleep, two whilst supine. The sleep latency was **42.5 min** and the sleep efficiency was **81.8%.** He felt he didn't sleep well.

Respiratory Events:
There were moderate hypopnoeas observed throughout the study, with a total RDI of = **29.4/hr**. Whilst in REM sleep, there were severe hypopneas observed at a rate of **65.5/hr**. These respiratory events were associated with an average SpO2 desaturation of **4 %,** and a nadir SpO2 of **83% .** Mild snoring observed.

Periodic Leg Movements: Rare PLMs observed. PLM Index = **2.8/hr.**

Cortical Arousal's: Arousal index = **9.7/hr.** Respiratory Arousal index = **6.3/hr.** PLM Arousal Index = 0.9/**hr**.

ECG abnormalities: Normal sinus rhythm.

Diagnosis/Treatment:

Severe REM related SAHS. Recommend CPAP.

Respiratory and Sleep Physician   Senior Sleep Scientist

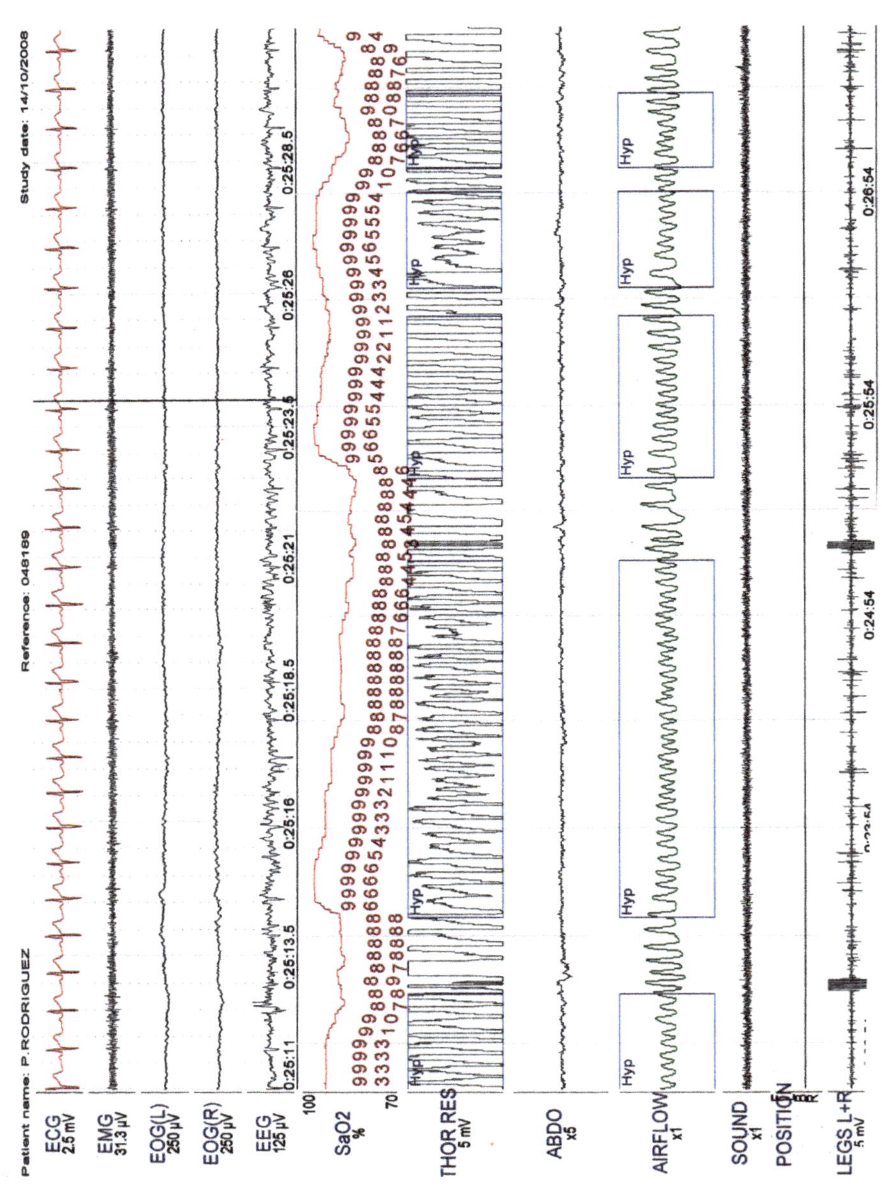
Patient name: P.RODRIGUEZ
Reference: 048189
Study date: 14/10/2008
ECG 2.5 mV
EMG 31.3 µV
EOG(L) 250 µV
EOG(R) 250 µV
EEG 125 µV
SaO2 %
THOR RES 5 mV
ABDO x5
AIRFLOW x1
SOUND x1
POSITION
LEGS L+R 5 mV
Hyp

Vaucluse Hospital. Sleep Disorders Centre.

Patient Name: Paul RODRIGUEZ
Date of Birth: 4/1/60
Weight: 95 kg
BMI: 31.0 kg/m^2

Study Date: 14/10/2008
Sex: M
Height: 175 cm
UR: 048189

SLEEP STATISTICS

Report time from 21:55:24 to 05:49:53			= 474.5 min
Time available for sleep (lights out)			= 474.5 min
Sleep latency			= 42.5 min
REM latency			= 98.5 min
Sleep period from 22:37:54 to 05:49:53			= 432.0 min
Total time awake during sleep period			= 44.0 min
Stage 1 = 19.5 min 5.0%		**Total Sleep**	**= 388.0 min**
Stage 2 = 222.0 min 57.2%		NREM Sleep	= 305.5 min 78.7%
Stage 3 = 57.5 min 14.8%		**REM Sleep**	**= 82.5 min 21.3%**
Stage 4 = 6.5 min 1.7%		Movement time	= 0.0 min
		Sleep Efficiency	= 81.8%

RESPIRATORY / SLEEP STATISTICS

	NREM			REM		
	Back	Other	All	Back	Other	All
SaO_2% min average	92	94	94	91	93	92
SaO_2% lowest	83	90	83	83	90	83
RDI						
Unsure	0.0	0.0	0.0	0.0	0.0	0.0
Central Apnea	0.0	0.0	0.0	0.0	0.0	0.0
Obstructive Apnea	0.0	0.0	0.0	2.4	0.0	1.5
Mixed Apnea	0.0	0.0	0.0	0.0	0.0	0.0
Hypopnea	70.8	14.0	19.6	59.4	70.9	64.0
Apnea+Hypopnea	70.8	14.0	19.6	61.8	70.9	65.5
			19.6			65.5

Total RDI	= 29.4
SaO2 awake average	= 95 %
Average SaO2 desaturation	= 4 %
Mean Apnea / Hypopnea duration	= 26.8 sec
Longest Apnea	= 29 sec
Longest Hypopnea	= 101 sec

AROUSAL STATISTICS (Sleep time)

	REM	NREM	Total
Per hour			
RESP	16.0	3.7	6.3
SPONT	0.7	2.9	2.5
PLM	0.0	1.2	0.9
			9.7

PLM STATISTICS (Sleep time)

Number of PLM per hour of NREM Sleep	= 3.5
Number of PLM per hour of REM Sleep	= 0.0
Number of PLM per hour of Sleep	= 2.8

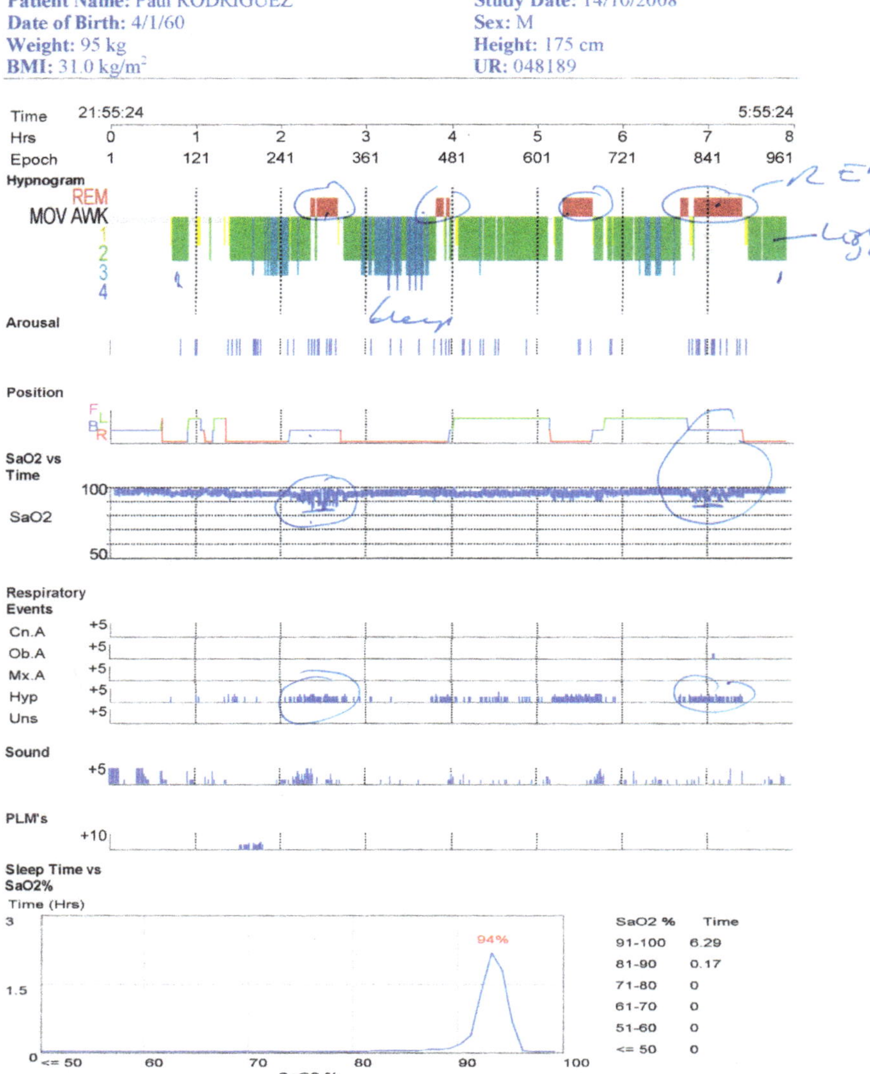
Vaucluse Hospital. Sleep Disorders Centre
Patient Name: Paul RODRIGUEZ
Date of Birth: 4/1/60
Weight: 95 kg
BMI: 31.0 kg/m²
Study Date: 14/10/2008
Sex: M
Height: 175 cm
UR: 048189
Time 21:55:24 5:55:24
Hrs 0 1 2 3 4 5 6 7 8
Epoch 1 121 241 361 481 601 721 841 961
Hypnogram
REM
MOV AWK
1
2
3
4
REM
Light sleep
Arousal
Position
F
L
B
R
SaO2 vs Time
SaO2
100
50
Respiratory Events
Cn.A
Ob.A
Mx.A
Hyp
Uns
+5
Sound
PLM's
+10
Sleep Time vs SaO2%
Time (Hrs)
3
1.5
0
94%
<= 50 60 70 80 90 100
SaO2 %
SaO2 % Time
91-100 6.29
81-90 0.17
71-80 0
61-70 0
51-60 0
<= 50 0

Appendix 2

Graphs of Weekly Averages for Control Pause and hours of Sleep/Nightly Wakes[21]

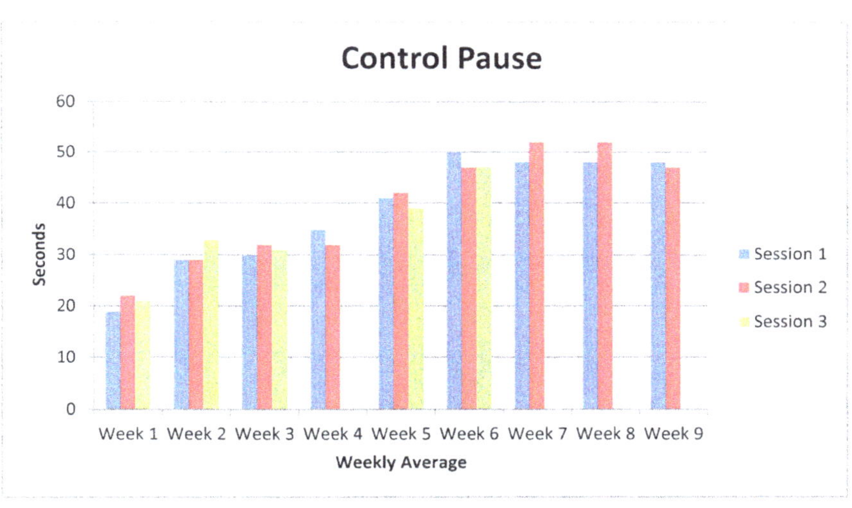

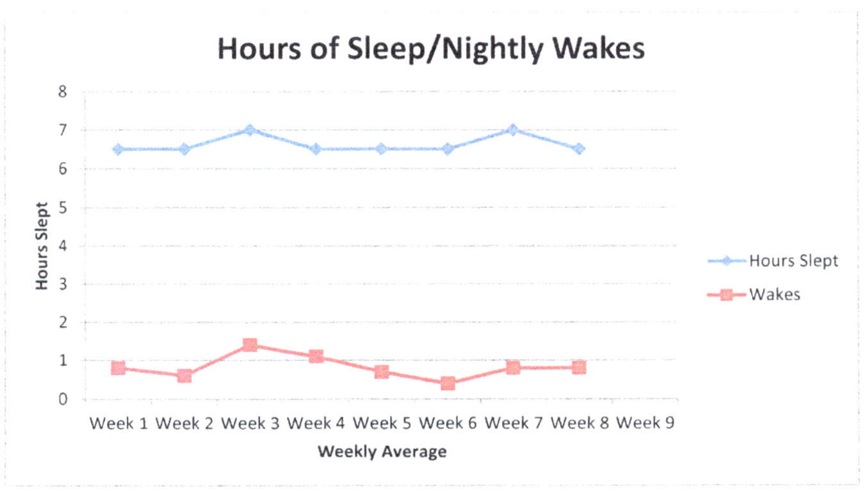

Appendix 3

Comparative photographs "before" and "after" of the Author[28]

14/08/2009

25/05/2010

Appendix 4

Sleep Study 20.05.2010[29]

ST VINCENTS & MERCY PRIVATE HOSPITAL
SLEEP DISORDERS CENTRE

St Vincents & Mercy
Private Hospital

Patient: Paul RODRIGUEZ **Study Date:** 20/05/2010

Study Type: DIAGNOSTIC

Interpreting Physician:

Date of Birth:	4/01/1960	Height:	177 cm	BMI:	27.8 kg/m^2
Gender:	Male	Weight:	87 kg		

TECHNOLOGIST'S COMMENTS:

1. **Technical Considerations:**
   1.1. Patient refused to wear nasal cannula for the entire study. Likely that respiratory disturbance is slightly under or over scored as a result.
   1.2. Patient had mouth taped as per Buteyko technique.

2. **Sleep Perception:** Paul's self-perceived sleep quality was worse than usual and was satisfactorily restorative. The patient reported sleeping for a total of 5 hours, with a latency of 15 minutes.

3. **Sleep Architecture:** Sleep was reasonable with slightly reduced SWS (17.2%) and good REM (18.4%). The sleep efficiency was 84.0%, with a latency of 5.0minutes (LPS 11.5mins) and a Total Sleep Time of 343.0minutes. There were three episodes of REM (first while partially supine), with a latency of 83.5minutes.
   3.1. Arousals: Arousals were moderately elevated in frequency & predominantly spontaneous in nature. Respiratory AI = 2.1/hr. PLM AI = 7.2/hr. **Total AI = 44.8/hr.**
   3.2. Posture: Body position was mostly LATERAL throughout the sleep period, with supine sleep comprising 22% of TST (78.0mins).

4. **Respiratory Disturbance:** There were infrequent hypopnoeas observed throughout mostly during REM. Central apnoeas were observed at sleep onset times and post gross body movement.
   AHI_{TOTAL} = 1.6/hr. AHI_{REM} = 3.8/hr. AHI_{NREM} = 1.1/hr
   RDI_{TOTAL} = 10.8/hr. RDI_{REM} = 13.3 /hr. RDI_{NREM} = 10.3/hr.
   Respiratory-related arousal was associated with 100% of all scored events. Oxygen desaturation was minimal, with SaO_2 dropping to a Nadir of 93% (NREM=93%, REM=93%), from an awake baseline of 95%.
   4.1. RERAs: COMMENT $RERA_{TOTAL}$ =9.2/hr. $RERA_{REM}$ =9.5/hr. $RERA_{NREM}$ = 9.2/hr
   4.2. Mild snoring was present mostly during SUPINE SWS however a few snores were observed upon arousal during non-supine REM. Average snoring intensity was 52.9 dB.

5. **Limb Movements:**
   5.1. Periodic Limb Movements during sleep were present during first cycle of NREM. PLMs were also observed during REM however these are possibly due to an underscoring of respiratory events and RERAs. **PLM Index = 11.2/hr. PLM AI = 7.2/hr.**
   5.2. Leg movements during sleep were infrequent outside of PLM sequences and were mostly secondary to arousals. Leg Movements = 19.6/hr (NREM = 16.9/hr; REM = 31.4/hr).
   5.3. Leg movements were reasonably frequent during wake prior to lights out.

6. **ECG Rhythm:** Normal Sinus Rhythm was noted. The average heart rate was 61 bpm.

Sleep Technologist

ST VINCENTS & MERCY PRIVATE HOSPITAL
SLEEP DISORDERS CENTRE

DIAGNOSTIC SLEEP STUDY

SLEEP STATISTICS

Start Recording Time	= 22:46:48	Total Sleep Time	= 343.0 min
End Recording Time	= 06:02:00		
Time in bed	= 408.5 min	WASO	= 60.5 min
Time available for sleep (lights out)	= 408.5 min		
		Wake (min)	= 65.5 min
Lights Out Time	= 23:13:18	Stage 1 (min)	= 34.0 (9.9% TST)
Lights On Time	= 06:01:47	Stage 2 (min)	= 187.0 (54.5% TST)
		Stage 3/4 (min)	= 59.0 (17.2% TST)
Sleep latency	= 5.0 min	**NREM (min)**	**= 280.0 (81.6% TST)**
Latency to persistent sleep	= 11.5 min	**REM (min)**	**= 63.0 (18.4% TST)**
REM latency	**= 83.5 min**	**Sleep Efficiency**	**= 84.0%**

	NREM			**REM**			**ALL SLEEP**		
	Supine	Other	All	Supine	Other	All	Supine	Other	All
AROUSALS									
Respiratory #			6			6			12
RERA #			41			6			47
PLM #			40			1			41
Spontaneous #			124			32			156
Respiratory (/hr)			1.3			5.7			2.1
RERA (/hr)			8.8			5.7			8.2
PLM (/hr)			8.6			1.0			7.2
Spontaneous (/hr)			26.6			30.5			27.3
							TOTAL		44.8 /hr
POSITION									
Time spent (mins)	68.5	211.5	280.0	9.5	53.5	63.0	78.0	265.0	343.0
RESPIRATORY									
Hypopnea #	0	3	3	0	4	4	0	7	7
Obstructive #	0	0	0	0	0	0	0	0	0
Central #	0	2	2	0	0	0	0	2	2
Mixed #	0	0	0	0	0	0	0	0	0
Apnea + Hypopnea #	0	5	5	0	4	4	0	9	9
Hypopnea (/hr)	0.0	0.9	0.6	0.0	4.5	3.8	0.0	1.6	1.2
Obstructive (/hr)	0.0	0.0	0.0	0.0	0.0	0.0	0.0	0.0	0.0
Central (/hr)	0.0	0.6	0.4	0.0	0.0	0.0	0.0	0.5	0.3
Mixed (/hr)	0.0	0.0	0.0	0.0	0.0	0.0	0.0	0.0	0.0
Apnea + Hypopnea (/hr)	0.0	1.4	**1.1**	0.0	4.5	**3.8**	0.0	2.0	**1.6**
							TOTAL AHI		1.6/hr
Non-arousing events (/hr)			0.0			0.0			**0.0**
RERA (/hr)			9.2			9.5			**9.2**
RDI (/hr)			**10.3**			**13.3**			**10.8**
							TOTAL RDI		10.8/hr
SaO2 STATISTICS									
% of total desaturations between:									
0-2% inclusive			86			67			80
3-4% inclusive			14			33			20
≥ 5%			0			0			0
SaO_2% min avg	95	96	96	96	96	96	95	96	96
SaO_2% lowest	94	93	93	95	93	93	94	93	93

SaO_2 awake average = 96 %
Average SaO_2 desaturation = 2 %

SNORING STATISTICS

Average Snoring level = 52.9 dB
Maximum Snoring level = 63.1 dB

ST VINCENTS & MERCY PRIVATE HOSPITAL
SLEEP DISORDERS CENTRE

	WAKE	NREM	REM	TOTAL SLEEP
LIMB & PLM STATISTICS				
Limb Movements (/hr)	30.2	16.9	31.4	**19.6**
PLM (/hr)	0.0	10.1	16.2	11.2
PLM AI (/hr)		8.6	1.0	7.2
Proportion of PLMs associated with arousal (%)		**85**	**6**	**64**
HEART RATE				
Average heart rate		61	62	61
Minimum heart rate		50	48	48
Maximum heart rate		78	96	96

PLM Histogram
The Number of PLMs vs. Inter-event Interval in seconds

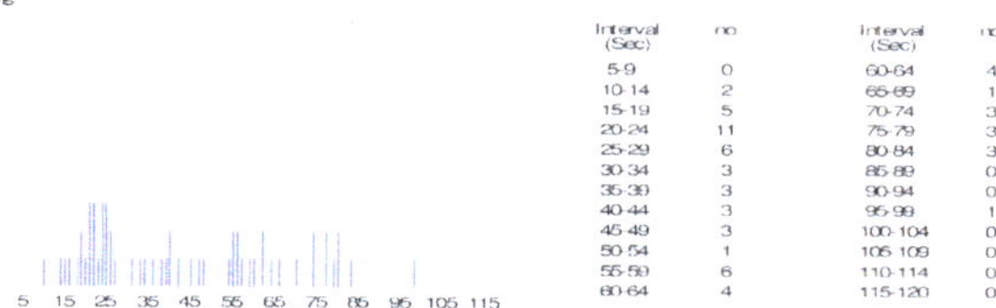

Interval (Sec)	no	Interval (Sec)	no
5-9	0	60-64	4
10-14	2	65-69	1
15-19	5	70-74	3
20-24	11	75-79	3
25-29	6	80-84	3
30-34	3	85-89	0
35-39	3	90-94	0
40-44	3	95-99	1
45-49	3	100-104	0
50-54	1	105-109	0
55-59	6	110-114	0
60-64	4	115-120	0

Heart Rate Histogram
Percentage Report Time vs. BPM

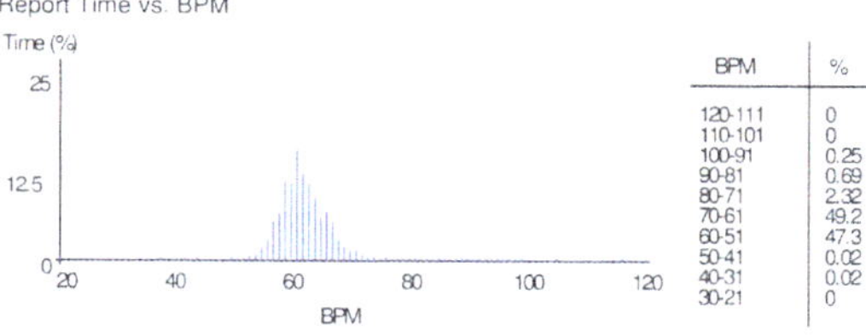

BPM	%
120-111	0
110-101	0
100-91	0.25
90-81	0.69
80-71	2.32
70-61	49.2
60-51	47.3
50-41	0.02
40-31	0.02
30-21	0

ST VINCENTS & MERCY PRIVATE HOSPITAL

SLEEP DISORDERS CENTRE

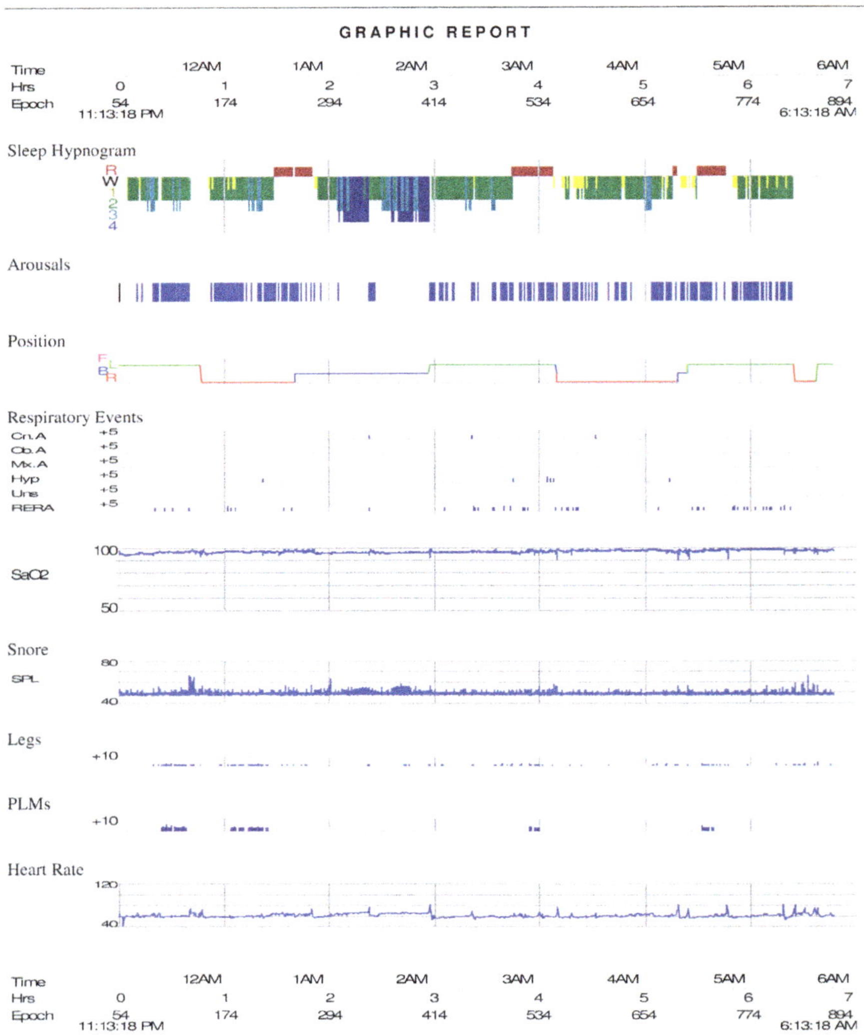

ST VINCENTS & MERCY PRIVATE HOSPITAL
SLEEP DISORDERS CENTRE

DEFINITIONS

Procedure
Attended polysomnography on Compumedics equipment. International 10-20 electrode placement. Recording EEG (central & occipital derivations), EOG, sub-mental EMG, ECG, airflow (thermistor & cannula), respiratory effort, oximetry, snoring (dB sound meter), body position, pulse rate, leg EMG & digital video. Treatment PAP, MAS, & supplemental Oxygen as requested by consultant.

Sleep Staging
Sleep staging is performed in accordance with the criteria of the Rechtschaffen & Kales Manual.

Arousal Rule
An arousal is scored during sleep stages 1, 2, 3, 4, or REM if there is an abrupt shift of EEG frequency including alpha, theta and / or frequencies greater than 16Hz (but not spindles) that lasts at least 3 seconds, with at least 10 seconds of stable sleep preceding the change. Scoring of arousal during REM requires a concurrent increase in submental EMG lasting at least 1 second.

Scored arousals are sub-classified into Respiratory, RERA, PLM or Spontaneous in accordance with their identified origin.

Respiratory Events

Apnoea
An apnoea is defined as a ≥80% drop from the baseline breathing amplitude for both airflow and nasal pressure, for a duration of ≥10 seconds, regardless of arousal or desaturation.

Obstructive Apnoea: An event which meets apnoea criteria and is associated with continued or increased inspiratory effort throughout the entire period of absent airflow.

Central Apnoea: An event which meets apnoea criteria and is associated with absent inspiratory effort throughout the entire period of absent airflow.

Mixed Apnoea: An event which meets apnoea criteria and is associated with an initial period of absent inspiratory effort, followed by resumption of inspiratory effort in the second portion of the event, throughout the period of absent airflow.

Hypopnoea
A hypopnoea is scored if all of the following criteria are met:

1. The nasal pressure signal excursions drop by ≥50% of the baseline breathing amplitude
2. The duration of this drop occurs for a period lasting at least 10 seconds
3. There is a ≥3% desaturation from pre-event baseline or the event is associated with arousal
4. At least 90% of the event's duration must meet the amplitude reduction criteria for a hypopnoea.

Apnoea & Hypopnoea Index (AHI): The sum of all apnoeas & hypopnoeas indexed per hour of sleep.

Respiratory Effort-Related Arousal (RERA)
A RERA is defined as an arousal preceded by any of the following:

1. An apnoea or hypopnoea of <10seconds in duration
2. The nasal pressure signal excursions drop by 20-50% of the baseline breathing amplitude, for a duration of ≥10seconds
3. A sequence of breaths lasting ≥10seconds characterized by increasing respiratory effort or flattening of the nasal pressure waveform, when the sequence of breaths does not meet the criteria for an apnoea or hypopnoea.

Respiratory Disturbance Index (RDI): All arousing events plus RERAs indexed per hour of Sleep.

Limb Movements & Periodic Limb Movements
The following rules define a significant limb Movement (LM):

1. The duration of a LM is 0.5 – 10 seconds
2. The LM must be clearly distinguishable from the background EMG, minimal amplitude criteria is not applicable.
3. A LM is not scored if it occurs during a period from 0.5seconds preceding a respiratory event to 0.5seconds following a respiratory event.

The following rules define a Periodic Limb Movement (PLM) series:

1. The minimum number of consecutive LMs needed to define a PLM series is 4LMs.
2. The period length between LMs to include them as part of a PLM series is 5 – 90seconds.
3. An arousal and a PLM are considered to be associated with each other when there is ≤3.0seconds between the end of one event and the onset of the other event regardless of which is first.

References:

Rechtschaffen A, Kales A. A Manual of Standardized Terminology, Techniques and Scoring System for Sleep Stages of Human Subjects. US Department of Health, Education, and Welfare Public Health Service – NIH/NIND; 1968

The AASM Manual for the Scoring of Sleep and Associated Events – Rules, Terminology and Technical Specifications. American Academy of Sleep Medicine: 2007.

Appendix 5

Professor Buteyko's Breathing Chart of Control Pause and Pulse Rate with alveolar CO_2 level and patient health[63]

Pulmonary Ventilation Criteria

Condition of the Body	Type of Breathing	Extent of Dysfunction	Alveolar CO_2 %	Alveolar CO_2 mmHg	Control Pause (CP) (sec)	Maximum Pause (MP) (sec)	Pulse per minute thebreathingman.com
Super Endurance	Superficial	V	7,5	54	180	210	48
		IV	7,4	53	150	190	50
		III	7,3	52	120	170	52
		II	7,1	51	100	150	55
		I	6,8	48	80	120	57
Normal			6,5	46	60	90	68
Disease	Deep	I	6,0	43	50	75	65
		II	5,5	40	40	60	70
		III	5,0	36	30	50	75
		IV	4,5	32	20	40	80
		V	4,0	28	10	20	90
		VI	3,5	24	5	10	100
		VII			DEATH		

Glossary

AHI	Apnoea Hypopnea Index
BBM	Buteyko Breathing Method
BMI	Body Mass Index
CAE	Council of Adult Education
CPAP	Continuous Positive Air Pressure
CO_2	Carbon Dioxide
CP	Control Pause
EDS	Excessive Daytime Sleepiness
ENT	Ear Nose & Throat
EP	Extended Pause
NO	Nitric Oxide
O_2	Oxygen
RBS	Reduced Breathing Session
REM	Rapid Eye Movement
RLS	Restless Legs syndrome
SaO_2	Symbol for percentage of available haemoglobin that is saturated with oxygen

Endnotes

Introduction

1. NIH/National Heart, Lung and Blood Institute 2009, "Severe sleep apnea tied to increased risk of death", *Science Daily*, 20 August, http://www.sciencedaily.com/releases/2009/08/090817190646.htm. Accessed 14 May 2013.

2. *Sleepdex – resources for better sleep* website, last updated 27 April 2014, http://www.sleepdex.org/epidemiologyapnea.htm. Accessed 8 September 2013.

3. The Secrets of Sleep by Dr Jerome Groopman The New Yorker 23 October 2017 https://www.newyorker.com/magazine/2017/10/23/the-secrets-of-sleep Accessed 6 March 2018

4. Deloitte Access Economics 2012, *Reawakening the nation: the economic cost of sleep disorders in Australia, Sleep Health Foundation, October 2011*, Deloitte Access Economics report for the Sleep Health Foundation, http://www.sleephealthfoundation.org.au/component/content/article.html?id=76:research. Accessed 23 January 2014.

5. Deloitte Access Economics 2017, Asleep on the job. Costs of inadequate sleep in Australia, Sleep Health Foundation, August 2017, Deloitte Access Economics report for the Sleep Health Foundation, https://www.sleephealthfoundation.org.au/86-sleep-health-foundation/admin/special-reports/867-asleep-on-the-job Accessed 7 March 2018.

6. "Feeling of uneasiness and restlessness in the legs after going to bed (sometimes causing insomnia)." *Advanced English dictionary*, version 6.2 © 2012.

Chapter One

7. *The Reader's Digest Oxford complete wordfinder,* 1994 edn, p. 61.

8. Stark, J & Stark, R 2002, *The carbon dioxide syndrome,* Buteyko On Line Ltd, Coorparoo, Queensland, p. 21.

9. Desai, AV, Grunstein, RR, Ellis E & Wheatley, JR 2003, "Fatal distraction: a case series of fatal fall-asleep road accidents and their medicolegal outcomes", *Medical Journal of Australia,* vol. 178 (8), pp. 396–399.

10. "A brief period of sleep, usually of a few seconds, that may result from sleep deprivation or various medical conditions." *Advanced English dictionary,* version 6.2 © 2012.

11. Desai, AV, Grunstein, RR, Ellis E & Wheatley, JR 2003, "Fatal distraction: a case series of fatal fall-asleep road accidents and their medicolegal outcomes", *Medical Journal of Australia,* vol. 178 (8), pp. 396–399.

12. Graham, T 2012, *Relief from snoring and sleep apnoea,* Penguin Group (Australia), Melbourne, p. 63.

13. Jeon, Y. J., Yoon, D. W., Han, D. H., Won, T.-B., Kim, D.-Y. and Shin, H.-W. (2015), *Low Quality of Life and Depressive Symptoms as an Independent Risk Factor for Erectile Dysfunction in Patients with Obstructive Sleep Apnea.* J Sex Med, 12 http://onlinelibrary.wiley.com/doi/10.1111/jsm.13021/full Accesed 7 March 2018

14. Ohayon Maurice, MD, PhD, 2003 *Study finds link between sleep apnea, depression. A breathing-related sleep problem is more likely to occur in depressed patients* Journal of Clinical Psychiatry referred to in the Stanford Report, Stanford University 5 November 2003 https://news.stanford.edu/news/2003/november5/depression.html Accessed 7 March 2018)

Chapter Two

15. *Buteyko Health & Breathing Newsletter,* vol. 10, 2008/9, p. 2.

16. See Appendix 1 Diagnostic Sleep Study Report 14 October 2008. Vaucluse Hospital, Brunswick.
17. Continuous Positive Air Pressure.
18. I later discuss what is known as the "Bohr effect" in Chapter 7 of this book.
19. Article 22 of the Third Geneva Convention provides that a prisoner of war cannot be held in conditions that are prejudicial to their health.
20. Stark, J & Stark, R 2002, *The carbon dioxide syndrome*, Buteyko On Line Ltd, Coorparoo, Queensland, p. 138.
21. See Appendix 2 Weekly Average for CP and Hours of Sleep/Nightly Wakes over 9 Week Period.

Chapter Three

22. "A partition of bone and cartilage between the nasal cavities." *Advanced English dictionary*, version 6.2 © 2012.
23. Graham, T 2012, *Relief from snoring and sleep apnoea*, Penguin Group (Australia), Melbourne, p. 49.
24. Cole, P & Haight JS 1984, "Posture and nasal patency", *American Review of Respiratory Diseases*, vol. 129, pp. 351–354, referred to in J Stark & R Stark 2002, *The carbon dioxide syndrome*, Buteyko On Line Ltd, Coorparoo, Queensland, p. 120.
25. Davies, AM, Koenig, JS & Thach BT 1989, "Characteristics of upper airway chemoreflex prolonged apnea in human infants", *American Review of Respiratory Diseases*, vol. 139, pp. 668–673, referred to in J Stark & R Stark 2002, *The carbon dioxide syndrome*, Buteyko On Line Ltd, Coorparoo, Queensland, p. 120.
26. Graham, T 2012, *Relief from snoring and sleep apnoea*, Penguin Group (Australia), Melbourne, p. 57. Reproduced with permission from Penguin Group (Australia), Melbourne.
27. Graham, T 2012, *Relief from snoring and sleep apnoea,* Penguin Group (Australia), Melbourne, p. 179.

Chapter Four

28. See Appendix 3 Comparative Photographs of the Author "Before" (14.08.2009) and "After" (25.05.2010).

29. See Appendix 4 Diagnostic Sleep Study Report 20 May 2010 (published by St Vincent's & Mercy Hospital).

30. Chung, F, Liao, P et al. 2012, "Oxygen desaturation index from nocturnal oximetry: a sensitive and specific tool to detect sleep disordered breathing in surgical patients", *Anesthesia & Analgesia*, vol. 114, no. 5, p. 994.

31. "Polysomnography", *MedlinePlus* website (a medical encyclopedia), 31.7.2011, http://www.nlm.nih.gov/medlineplus/ency/article/003932.htm. Accessed 4 June 2013.

32. Definitions of "obstructive apnoea" and "central apnoea" in *Diagnostic sleep study report 20 May 2010*, St Vincent's & Mercy Hospital.

33. Deloitte Access Economics 2017, *Asleep on the job. Costs of inadequate sleep in Australia, Sleep Health Foundation, August 2017*, Deloitte Access Economics report for the Sleep Health Foundation, https://www.sleephealthfoundation.org.au/86-sleep-health-foundation/admin/special-reports/867-asleep-on-the-job Accessed 12 March 2018.

34. "A decrease in the level of oxygen saturation in a patient's haemoglobin." *Advanced English dictionary*, version 6.2 © 2012.

35. Letter from my sleep specialist to my GP dated 2 June 2010.

36. Professor Peter Eastwood, Director of The University of WA Centre for Sleep Science, speaking on Richard Adey's Radio National program *Life Matters* on 10 June 2011.

37. Professor Peter Eastwood quoting an Access Economics Report circa 2004 on Radio National program *Life Matters* on 10 June 2011.

38. Birch, M 2012, *Sleep apnoea and breathing retraining*, Buteyko Institute of Breathing & Health, Melbourne, pp. 1–20. Report by Mary

Birch, registered nurse and breathing retraining consultant, on behalf of BIBH.

39. Birch, M 2012, *Sleep apnoea and breathing retraining*, Buteyko Institute of Breathing & Health, Melbourne, p. 5. Report by Mary Birch, registered nurse and breathing retraining consultant, on behalf of BIBH.
40. Birch, M 2012, *Sleep apnoea and breathing retraining*, Buteyko Institute of Breathing & Health, Melbourne, p. 3. Report by Mary Birch, registered nurse and breathing retraining consultant, on behalf of BIBH.
41. *Medical Journal of Australia* (December 1998) as reported on website of Buteyko Health & Breathing, http://www.sleepingallnight.com/scientific.htm. Accessed 28 September 2014.

Chapter Five

42. A tribute to Professor Konstantin P Buteyko on his death, from Leo Volkov, Buteyko practitioner, http://www.members.westnet.com.au/pkolb/trib_LV.htm. Accessed 28 September 2014.
43. "A non-steroidal anti-inflammatory and analgesic medicine." *Advanced English dictionary*, version 6.2 © 2012.
44. Altukhov, S 2009 *Dr Buteyko's discovery: volume 1 – the destruction of the laboratory*, English translation by M Farquharson, F Paterson, H Stacey & D Steele, published by Alex Spence, p. 36. Originally published in Russian in 1993 by Altmilla Publishing House, Novosibirsk.
45. *Altukhov, S 2009 Dr Buteyko's discovery: volume 1 – the destruction of the laboratory,* English translation by M Farquharson, F Paterson, H Stacey & D Steele, published by Alex Spence, p. 36. Originally published in Russian in 1993 by Altmilla Publishing House, Novosibirsk.
46. Graham, T 2012, *Relief from snoring and sleep apnoea*, Penguin Group (Australia), Melbourne, p. 61.

47. American Thoracic Society 2005, "Moderate to severe sleep apnea significantly raises stroke risk, study finds", *Science Daily*, 23 May, http://www.sciencedaily.com/releases/2005/05/050523153827.htm. Accessed 14 May 2013.

48. American Thoracic Society, 2008, "Obstructive sleep apnea causes earlier death in stroke patients, study finds", *Science Daily*, 20 May, http://www.sciencedaily.com/releases/2008/05/080518182655.htm. Accessed 14 May 2013.

49. Graham, T 2012, *Relief from snoring and sleep apnoea*, Penguin Group (Australia), Melbourne, p. 188.

50. Quote by the Chinese philosopher Confucius (551–479 BC).

51. Rinpoche, A & Choying Zangmo, A 2013, *The Tibetan yoga of breath,* Shambhala Publications Inc., Boston, Massachusetts, p. 100.

Chapter Six

52. *Orofacial Myofunctional Therapy: The Critical Missing Element to Complete Patient Care*, by Joy L Moeller, RDH, BS, COM http://www.Joymoeller.com Accessed 25 January 2017

53. *A frequent phenotype for paediatric sleep apnoea: short lingual frenulum* by Christian Guilleminault, Shehlanoor Huseni & Lauren Lo, https://www.ncbi.nlm.nih.gov/pmc/articles/PMC5034598/ Accessed 9 January 2018

54. *Is it Mental or is it Dental?* Raymond Silkman, DDS March 29 2006 Weston Price Foundation

55. Advanced English Dictionary Version 11.0 © 2018

56. Advanced English Dictionary Version 11.0 © 2018

57. *How Eating More Slowly can help you lose weigh*t by Franziska Spritzler, RD, CDE 1 February 2016 https://www.healthline.com/nutrition/eating-slowly-and-weight-loss Accessed 9 January 2018

58. *Comments made by Dr Lauren Boundy, Osteopath, B.Sci (Clin sci) M.H.Sci (Osteo) in an email to the author dated 8 February 2017*

Chapter Seven

59. *The Carbon Dioxide Syndrome* (2002) Jennifer Stark and Russel Stark at p 17
60. *Behavioural and Psychological Approaches to Breathing Disorders* K Naifeh (1994) at p 17 as referred to in The Carbon Dioxide Syndrome (2002) Jennifer Stark and Russel Stark at p 17
61. *The Carbon Dioxide Syndrome* (2002) Jennifer Stark and Russel Stark at p 26
62. *Human Physiology and Mechanisms of Disease* AC Guyton (1982) pp 279-80, 285, 300, 306, 320 - 323 as referred to in *The Carbon Dioxide Syndrome* (2002) Jennifer Stark and Russel Stark at p 22
63. See Appendix 5 – *Professor Buteyko's Breathing Chart correlating control pause and pulse rate with alveolar CO2 level and patient health retrieved from www.thebreathingman.com*
64. Metabolic equilibrium actively maintained by several complex biological mechanisms that operate via the autonomic nervous system to offset disrupting changes - *Advanced English Dictionary version 6.2* © 2012
65. *Carbon Dioxide* Yandell Henderson (1922), New Haven, Connecticut
66. *Nunn's Applied Respiratory Physiology* AB Lumb (2000) at pp 3-12, 237,267, 321, 362 as referred to in *The Carbon Dioxide Syndrome* (2002) Jennifer Stark and Russel Stark at p 31
67. *The Reader's Digest Oxford Complete Wordfinder* (1994 ed) p 672
68. *Nunn's Applied Respiratory Physiology* AB Lumb (2000) at pp 3-12, 237,267, 321, 362 as referred to in *The Carbon Dioxide Syndrome* (2002) Jennifer Stark and Russel Stark at p 31
69. *The Carbon Dioxide Syndrome* (2002) Jennifer Stark and Russel Stark at p 36
70. *The Carbon Dioxide Syndrome* (2002) Jennifer Stark and Russel Stark at p 36

71. *The Carbon Dioxide Syndrome* (2002) Jennifer Stark and Russel Stark at p 36

72. *The Psychology and Physiology of Breathing* (1993) R Fried at p 104 as referred to in *The Carbon Dioxide Syndrome* (2002) Jennifer Stark and Russel Stark at p 38

73. Jorissen M, Lefevere L, Willems T "Nasal Nitric Oxide" *Allergy* 2001: 56: pp 1026 – 1033

74. "virility drug (trade name Viagra) used to treat erectile dysfunction in men". *Advanced English dictionary*, version 11.0 © 2018

75. https://www.medicalnewstoday.com/articles/292292.php (Accessed 13 January 2018)

76. Lundberg, J O N and Weitzberg, E, 1999, "Nasal nitric oxide in man" *Thorax* pp 949-950

77. Lundberg, J O N and Weitzberg, E, 1999, "Nasal nitric oxide in man" *Thorax* pp 948

78. J Lundberg," Airborne nitric oxide: Inflammatory marker and aerocrine messenger in man" *Acta Physiologica Scandinavica,* Vol 157 no. 633 pp. 1-27, *1996* as referred to in De Sousa Michels et al. "Nasal Involvement in Obstructive Sleep Apnea Syndrome" *International Journal of Otolaryngology* Vol 2014 (2014), Article ID 717419 p 2

79. Bartley J and Wong C "Nasal Pulmonary Interactions" *Nasal Physiology and Pathophysiology of Nasal Disorders* 2013 pp 560 and 564

80. Djupesland Per G, Chatkin JM, Qian W, Haight JSJ, 2001 " Nitric oxide in the nasal airway: a new dimension in otorhinolaryngology" *Am J Otolaryngol* 22: 19-32 as referred to in Serrano C, Valero A, Picado C 2004 "Review Nasal Nitric Oxide" *Arch Bronconeumol* 40(5):222-30

81. Dusek J et al, 2005, "association between oxygen consumption and nitric oxide production during the relaxation response" https://www.ncbi.nlm.nih.gov/pubmed/16369463 Accessed 14 January 2018

82. Lundberg JO, Settergreen et al "Inhalation of nasally derived nitric oxide modulates pulmonary function in humans", *Acta Physiol Scand* 158: 343-347,1996

83. Crespo AS, Hallberg J, Lundberg JO et al, 2010 "Nasal nitric oxide and regulation of human pulmonary blood flow in the upright position", *Journal. Applied Physiology* 108: p 181

84. Crespo AS, Hallberg J, Lundberg JO et al,2010 "Nasal nitric oxide and regulation of human pulmonary blood flow in the upright position", *Journal. Applied Physiology* 108: p 187

85. Mosby's Medical & Nursing dictionary, 2nd edition, CV Mosby Company, USA, 1986 as referred to on page 5 of the BIBH report

86. *Report by Mary Birch, Registered Nurse & Breathing Retraining Consultant on behalf of BIBH, 2012* at p 5

87. *Op.Cit.at* p 5

88. A state in which the level of carbon dioxide in the blood is lower than normal; can result from deep or rapid breathing - - *Advanced English Dictionary version 6.2 © 2012*

89. Meah,MS, Gardner, WN. Post-hyperventilation apnea in conscious humans. J Physiol, 1994; 477:527-538 as referred to in *Report by Mary Birch, Registered Nurse & Breathing Retraining Consultant on behalf of BIBH, 2012 at p 6*

90. *Relief from snoring and sleep apnoea* (2012) Tess Graham at p 51

91. systemic lupus erythematosus - an inflammatory disease of connective tissue with features including fever, weakness, fatigability, joint pains and skin lesions - *Advanced English Dictionary version 6.2 © 2012*

92. This phrase was first found in *Andrea del Sarto*, 1855, a poem by Robert Browning

Index

H

I

M

N

O

P

R

S

T

V

www.ingramcontent.com/pod-product-compliance
Lightning Source LLC
Chambersburg PA
CBHW042247040426
42334CB00044B/3082

* 9 7 8 0 9 9 4 1 6 9 4 0 2 *